A CAMERA TRAPPER'S *Companion*

A CAMERA TRAPPER'S *Companion*

An Introduction to Exploring Nature with Trail Cameras

CHRIS WEMMER

BRG SCIENTIFIC

WASHINGTON, DISTRICT OF COLUMBIA

A Camera Trapper's Companion: An Introduction to Exploring Nature with Trail Cameras

For rights and permissions, please contact: BRG.Scientific@gmail.com or

BRG Scientific LLC

3025 Ontario RD NW #407

Washington, District of Columbia 20009

www.BRGScientific.com

First edition

ISBN: 979-8-9862821-2-1 (Hard Cover)

ISBN: 979-8-9862821-4-5 (Paperback)

ISBN: 979-8-9862821-3-8 (Print Replica eBook)

Library of Congress Control Number: 2023916321

Editor: Melissa Stevens, Purple Ninja Editorial

Indexer: Shelley Quattrocchi, Quattro Indexing Service

Cover Design and Interior Formatting: Becky's Graphic Design® LLC

Dedicated to the memory of Professor Joseph G. Hall for opening an undergraduate's eyes to the allure and magic of camera trapping.

Note to the Reader: Glossary terms are bolded throughout. Endnotes for chapter sidebars are unnumbered and placed at the end of each chapter's notes at the back of the book.

Contents

CHAPTER 1

Starting the Journey

Camera trapping is on the rise. A growing number of people buy or make their own trail cameras, set them discreetly in the woods, and return to them later with as much anticipation as children on Christmas morning.

Why such growing enthusiasm? One reason is that camera trapping is a great way for people to satisfy their curiosity about nature and wildlife. The birders' tools of the trade don't help them see creatures that go bump in the night. Most mammals are nocturnal and secretive, and camera traps open their world to us. Though most camera trappers yearn for appealing photos, and some are quite skillful at getting them, pastoral outings and the thrill of discovery are almost always ample rewards. Simply said, camera trapping is an educational pursuit masquerading as a peaceful personal adventure in nature.

Camera trapping is also a relatively inexpensive form of wildlife photography. A camera trap costs a lot less than a high-end **DSLR** camera system, which is not to say that some camera trappers don't go overboard buying cameras and then have to answer to their spouse. The pursuit can certainly become addictive, and those who make their own cameras often get hooked on the personal satisfaction of tinkering.

Camera trapping is a boon to wildlife photography because camera trappers can log thousands of hours in the field using digital cameras and memory cards without actually spending hours and hours sitting in the field. Sit-and-wait wildlife photographers just can't afford that kind of field time, but they can use a camera trap to gather intelligence and, when their quarry's habits are known, return and sit it out for the perfect picture.

As a business phenomenon, the rise of the modern camera trap, or trail camera, is a remarkable story. Since the first commercial trail camera made its appearance in the late 1980s, sales have soared, manufacturers have multiplied, and the market has proliferated. The original target

Figure 1.1. Professor Joe Hall, in a leather field kilt, holds a Kaibab squirrel *(Sciurus aberti kaibabensis)* as it recovers from anesthesia and radio-collaring. (Kaibab Plateau, Arizona)
Courtesy of Betty Hall

consumer was the American hunter intent on scoping out the woods for monster bucks with enormous antlers, but wildlife biologists and nature lovers knew a good thing when they saw it, and they too became trail camera consumers.

A consumer public enthralled with electronic gadgetry welcomed the commercial trail camera with open arms. It was one more gadget to feed their addiction to electronica, and it also opened new possibilities for outdoor recreation. Using remote cameras to capture wildlife images captivates many more people than just biologists, and significant numbers of recreational naturalists have joined the growing family of users. Models of trail cameras quickly diversified, and some manufacturers have since developed products specifically for birders and wildlife biologists.

Camera traps have led to many significant discoveries in science, including new species. A recent informal survey on a camera trapping forum revealed that its members, many of whom are hunters, do not have a single-minded interest in trophy bucks and game meat for the freezer. Their goal is to photograph wildlife. Dedicated camera trappers soon learn that success is proportional to effort. Persistence pays off whether they use one camera or a fleet of them. Trail cameras are here to stay, their future is bright, and we can expect surprising improvements in their design and capability.

My awareness of camera traps emerged when I was a kid, in the 1950s. A morbid high-speed photo of a mouse execution in *Popular Mechanics* magazine made a lasting impression. The luckless rodent was airborne as the spring-bar of a mousetrap closed in for the kill. The possibility of photographing even a mouse with a remote camera intrigued me greatly; but in those days you had to make your own, and if your father was a music teacher, he probably was not the one to show you how.

It wasn't until college that I made my first camera trap, a doorbell-camera contraption, with guidance from the late Professor Joe Hall of what is now San Francisco State University (Figure 1.1). Joe taught the field classes—Ecology, Natural History of the Vertebrates, and Mammalogy—and was the only professor who lectured in jeans with a sports jacket and tie, donned a tartan kilt at special occasions, and wore a cowhide kilt for fieldwork. He also infected us with

Figure 1.2A. The Nikon FE camera trap we used to photograph the Sulawesi palm civet *(Macrogalidia musschenbroekii)* had a water-resistant plywood housing and hinged back panel for viewing.

whacky spoonerisms, like "in one swell foop." But I digress. Joe's camera trap trigger was a whisker switch wired to a battery-operated doorbell. Activating the switch made a little wooden bar press the shutter button. It was a wonderful gadget, but you had to visit your camera daily to hand-crank the film in case it had taken a picture.

Figure 1.2*B*. Team member Ken Lang tests the camera trap's electronic beam trigger (Sti 2090) with a machete. (Lore Lindu National Park, Sulawesi, Indonesia)

In those days camera trappers didn't fuss about such inconvenience. The remotely triggered camera was an excuse for budding naturalists to vicariously reenact one of the more benign aspects of frontier life—traipsing about the woods in search of animal spoor. Making the "**camera set**" was high adventure, though poison oak periodically laid them low. They were thrilled when they got a picture of a squirrel or deer, and along the way they developed observational skills, learned the local plants and animals, and exercised more caution when peeing in the woods.

In 1980 I found myself in faraway Indonesia with colleagues from the Smithsonian National Zoological Park, where I was employed at the time. Our quest was to find the little-known Sulawesi civet (*Macrogalidia musschenbroekii*). Camera technology was becoming electronic, and we were equipped with an improved homemade camera trap[1] (Figure 1.2*A*). The Nikon FE advanced the film after each exposure and was equipped with an outlet for a remote shutter device, which we cabled to an external photoelectric trigger. Setting the camera to trigger our elusive quarry was another matter. After positioning the clock and dangling a banana within the frame, the photoelectric beam had to be adroitly centered on an obscenely obvious reflector (Figure 1.2*B*). Only then was the set ready for action. Two months later we were back in the United States, impatiently awaiting our developed films. We were crestfallen when they arrived. Box after box contained slides of night-flying moths and falling leaves. But there were also a few photos of the mysterious civet, and that was enough to deem our enterprise a success (Figures 1.3*A* and *B*).

By the 1990s I was a full-blown zoo administrator subjected to harrowing commutes for endless meetings in Washington, DC. I wistfully observed my younger colleagues pursuing fieldwork with commercial trail cameras and vowed to someday indulge my passion for camera trapping in wild landscapes. Ever since I retired, I've been doing just that, and I've learned it isn't as simple as it seems. The novice camera trapper has to hone skills and climb a learning curve to reap satisfaction, and this requires patience. The first image of an animal, no

Figure 1.3*A*. As the civet seized the banana, its tail broke the photoelectric beam and shuttered the photo. (Lore Lindu National Park, Sulawesi, Indonesia; Nikon FE)

Figure 1.3*B*. After eating the bait, the civet stalked the camera. (Nikon FE)

matter how incomplete or out of focus, is a charm. But the charm soon wanes and expectations rise. The disenchanted retire their camera to the closet, while the willful push on, experiment with different sets and cameras, seek photos of different species, grow knowledge, and cultivate skills. For those who accept the challenge of possibilities, camera trapping becomes a compelling and fulfilling diversion.

Camera trapping has a certain mystique. It can teach us much about animals we rarely see in the flesh, and we may even get to know some of them as individuals. But our subjects know us largely by the clues we leave. They can certainly hear us in the distance—footfalls, snapping sticks, and barking dogs. They can certainly detect our visible or whiffy traces in the landscape. And they may even be able to gauge the recency of our visit, but do they know us as more than scattered clues? Now and then the camera trapper enters the twilight zone and makes a cognitive connection with one of "the others." The raven, bear, or mountain lion that showed up as soon as you left had surely been watching you set the camera. They are not oblivious, and they may know more about us than we think.

Camera trapping can be a high-stakes adventure. It is oddly thrilling to leave so much to chance or submit to risky temptations, like setting your camera afloat on a small raft in a marsh. Losing cameras and accessories to thieving humans, curious bears, or floodwaters is often a clear and present danger. But you hope for the best when you cast your camera's fate to the wind. Usually, misfortune takes a pass, and when you download your photos at home, you savor "the catch" with quiet satisfaction. The glow of discovery can last a week when serendipity delivers the unexpected image of a fox with a limp rabbit in its mouth or a video clip of bear cubs engaged in rough-and-tumble play. That's when you share your finding with friends and perhaps search the internet to learn more about your subjects.

Camera traps feed our curiosity about the natural world. Since their appearance in the commercial market, technical articles and books about methodology have surged, proving they are indispensable to wildlife biologists. But as educational tools for the lay public, they have not received equal treatment. Currently there are three inspired exceptions: Janet Pesaturo's *Camera Trapping Guide: Tracks, Sign, and Behavior of Eastern Wildlife*, Roland Kay's *Candid Creatures: How Camera Traps Reveal the Mysteries of Nature*, and Jack and Anna Mary Childs' *Ambushed on the Jaguar Trail, Hidden Cameras on the Mexican Border*. This volume is meant as a resource for anyone infected with the camera trapping bug—amateur naturalists, outdoor enthusiasts, citizen scientists, science teachers, and other educators.

My conviction is that camera traps are amazing tools for learning about nature. Using them can raise a lot of questions and lure you down multiple paths of study, but many folks need guidance and want to hear about the possibilities. The chapters that follow tell you how, starting with two scoops of camera trapping history. I've seasoned them with anecdotes, object lessons, and even some bedtime stories, and the documentation for each chapter can be found in the list of resources and endnotes.

Rapidly changing technology means sooner or later some of this information will go out of date, and internet links will eventually greet the reader with the disturbing message: Page not found. Technology has another worrisome consequence: the growing human appetite for any kind of virtual experience over real-time adventure in nature. Ironically, camera trapping uses technology for a noble purpose—the enjoyment, appreciation, and understanding of the natural world.

CHAPTER 2

Pioneers of Camera Trapping: 1880–1945

At the turn of the twentieth century, cameras were well suited for portraits and landscapes, but too cumbersome for stalking and photographing wild animals in nature. To photograph wildlife, naturalists built mechanical triggers that enabled animals to activate the shutter and ignite flash powder. These set, or **flash light**, cameras were ungainly, time-consuming to set up, and hard to conceal. When the flash powder ignited, the booming report terrified the animals and often woke the sleeping photographer. The resulting images captured the imagination of a growing nation that was becoming increasingly isolated from nature, but they told us little about the lives of their makers. Fortunately, several pioneers of camera trapping shared their adventures in writing and revealed their thrills of discovery, meditations on nature, and reflections on ethical issues.

THE DAWN OF CAMERA TRAPPING

In 1898 George Shiras III prepared to take his first picture of a white-tailed deer. He was afloat in a flat-bottomed skiff on Whitefish Lake (now Peter White Lake) in the upper peninsula of Michigan with a 5 × 7 plate camera fastened to the bow (Figure 2.1). The moment had arrived. He pulled the slide from the film holder and exposed the plate by briefly removing and replacing the lens cap. Or so he thought. He discovered later that he had pulled the slide from the wrong side of the film holder. He blamed the failure on buck fever, but he wasn't discouraged. He had just embarked on a journey of wildlife photography that lasted sixty-five years.

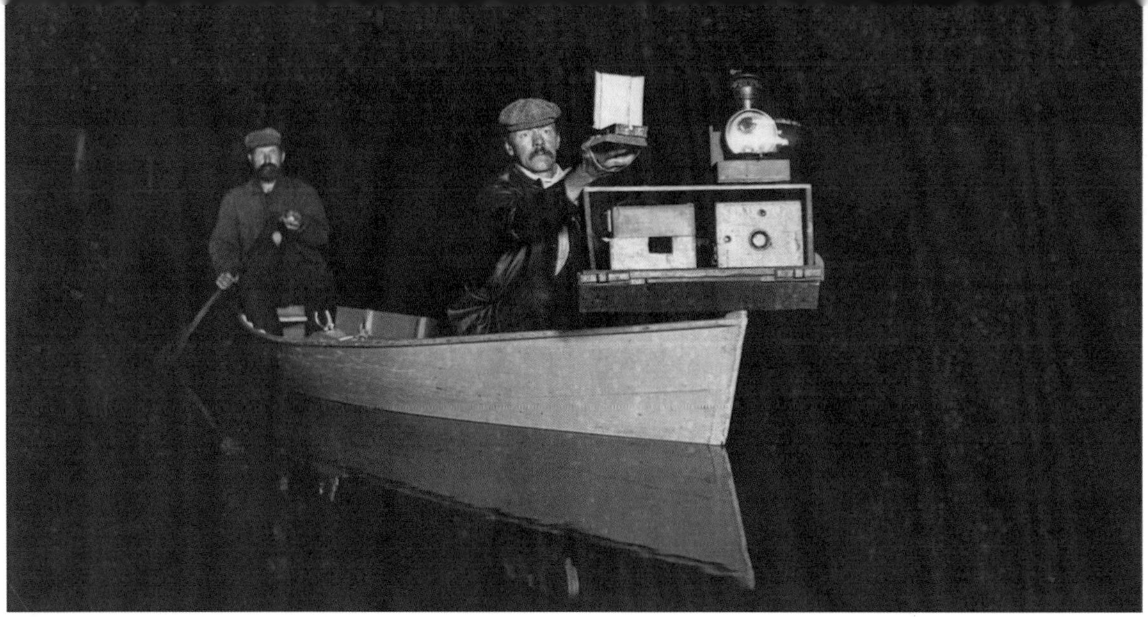

Figure 2.1. George Shiras III is seen at the helm camera-stalking nocturnal wildlife by skiff in Whitefish Lake, Michigan. *Courtesy of the National Geographic Society Library and Archives*

Shiras didn't plan on becoming a naturalist. His father sent him to private school in Massachusetts, far from the family home in Pittsburgh; then after studies at Cornell and Yale, he joined his father's law firm. When President Benjamin Harrison appointed Shiras's father to the Supreme Court, thirty-three-year-old Shiras took over the family firm. Following in his father's footsteps, he sought political office and was elected to the Pennsylvania state legislature and then the US Congress.

Shiras had the means to do whatever he wanted, and one of his loves was hunting. From the age of eleven, his rustic side had been nurtured by long vacations on the shores of Lake Superior at the family's second house in Marquette, Michigan. The Big Water was the frontier of the north woods, and Shiras learned to hunt, fish, and live off the land at Whitefish Lake under the guidance of Jack La Pete, an Ojibway guide of "comical expression" with one arm "deeply scarred and shriveled" from a bear attack, whose volubility he attributed to a possible Irish ancestor (Figure 2.2). Shiras wrote that the contrast between the smoke-filled Ohio Valley and the clean wilderness of northern Michigan caused him "to give up the exactions of a professional life that I might be free to interpret the laws of nature rather than those of man."[1] Thus he retired from structured routine and dedicated himself to the full-time pursuit of wildlife photography. He had the family connections, resources, and means to do it.

With trial-and-error tinkering in Marquette and nocturnal forays on Whitefish Lake, Shiras's photographic experiments led him through a succession of bulky, large-format cameras. His first remote camera required his presence to trigger it from a tree 300 feet away. The camera was hidden offshore at the edge of a slough, and when the deer was in the right place, he whistled to get its attention. He wrote: "The approach of the deer to the focus point of the camera gave all the tenseness of feeling provided by hunting with a rifle, but the cord was pulled instead of the trigger."[2]

His flashlight photos appeared in various magazines, and in 1900 he submitted a collection of deer flashlight photos to the World Exposition in Paris, where they won accolades and Shiras received a gold medal and diploma. Four years later his photos won another gold prize at the Louisiana Purchase Exposition in St. Louis. When Gilbert Grosvenor, the president of the National Geographic Society, saw Shiras's work he decided to devote an entire issue to it. *National Geographic Magazine* published "Hunting Wild Game with Flashlight and Camera" in 1906. Two members of the managing board, both geographers, were so disturbed by the society's new direction that they quit their jobs in protest.

Figure 2.2. Ojibway Jack La Pete served as a hunting guide to three generations of the Shiras family and was George's source of natural history wisdom. *Courtesy of the National Geographic Society Library and Archives*

Shiras soon realized that if people could take a self-portrait simply by pulling a string, a wild animal could be induced to do the same. The German invention of ***Blitzpulver***, or flash powder, was only a few years old. The fizzing fuse alerted the deer before the magnesium powder could ignite, and the unimproved powder was "useless on damp and windy nights."[3] When it worked, it burned slowly. And the flash, he learned, didn't freeze movement. A jerk of the deer's head left an image that looked like a decapitated animal.

But flash powder improved, and Shiras set about devising a mechanism that released the shutter and fired the flash cap at the same time. Soon he boasted a handheld flashlight that could be fired "with the quickness and certainty of a pistol."[4] Triumph came in 1893 when he finally captured a night flash of a yearling buck. In 1906 he wrote: "All difficulty in getting night pictures from a canoe disappeared."[5] The downside was that the explosion could also be seen in Marquette, Michigan, 20 miles from Whitefish Lake, and the dull boom could be heard at least a mile away.

Experimentation was not without its dangers. "At once came a tremendous explosion of the drier powder, and the damper portion gave forth a brilliant spluttering flame with a cloud of stifling smoke, compelling me to leap overboard in order to extinguish the blaze on my boots and later that in the boat."[6]

In the early 1890s, Shiras invented the first true camera trap, but he did not publish about it until 1913. It was automatic and required service after every exposure. He called the rigged **Schmidt Detective camera** (a 4 × 5-inch plate camera) a set camera, trap camera, or flash light

Figure 2.3. A camera set at a beaver lodge. Shiras spliced a separate image of the wood-crafted assembly of the camera, batteries, and trigger into this photo. *Courtesy of the National Geographic Society Library and Archives*

Figure 2.4. Shiras often rigged a second flash to capture spectacular escape reactions of deer.
Courtesy of the National Geographic Society Library and Archives

camera, and by trial and error, he perfected his technique with a variety of game (Figure 2.3). With salt and bait, he set about attracting various species to his sets, and he perfected the mechanical trigger.

Shiras knew that the flashlight's explosion caused most of his subjects to make dramatic exits and that his cameras failed to capture the action. The subjects were almost always seen peacefully tripping the wire. "I often wondered," he wrote, "just how an animal would react to the surprise of the flash."[7] He set two cameras with separate flashlight mechanisms, so that the firing of the first camera would trigger the second after a momentary delay. "The result obtained was sometimes extraordinary and always interesting," he wrote. Then he arranged the trip cord to fire a blank cartridge, which activated a drop weight to trigger a single camera. Among Shiras's most spectacular photos are the getaway shots of crouching, sprinting, and leaping deer (Figure 2.4).

IMITATION AND FLATTERY

As Shiras honed his photographic skills at the end of the nineteenth century, a wealthy German named Carl Georg Schillings was establishing his bona fides as a hunter–naturalist in German East Africa (Figure 2.5). Six years younger than Shiras, Schillings decried wanton killing of African game, but did not forsake rifles for cameras. A naturalist in the style of Teddy Roosevelt, Schillings wanted trophies for his personal museum and also donated specimens to Berlin's Zoological Museum. He captured live animals for the Berlin and London zoos and collected zoological curiosities, such as the ticks that circled the eyes of colobus monkeys. The parasite was new to science and was named *Ixodes schillingsi*.

Schillings made four trips to German East Africa between 1896 and 1903 and, after his first trip, began experimenting with flashlight photography and the trip camera. With encouragement from Berlin Zoo director Ludwig Heck, Schillings and army photographer Captain Kiesling worked on the camera at the factory of C. P. Goerz, the Berlin-based camera and lens manufacturer. The finished camera was used on at least two of the four expeditions, and the photographs were published in 1905 in a two-volume travel log titled *With Flash-Light and Rifle: Photographing by Flash-Light at Night the Wild Animal World of Equatorial Africa*.[8]

Shortly before Schillings's book was published, Shiras attended the Boone and Crockett Club's annual banquet with President Roosevelt and listened to the guest speaker, the German ambassador, praise the remarkable flashlight feats of his countryman Carl Georg Schillings. The ambassador was perplexed by the smiles in the audience, and Roosevelt told him it was the man sitting across from him, Mr. Shiras, who had developed camera trap methodology years ago and who had in fact won a gold medal for his photographic achievements at the 1900 World's Fair in Paris. Shiras listened politely to the interaction, but later searched his memory and recalled forgotten correspondence with a German professor from the Berlin Museum, most probably Professor Paul Matschie. A letter was still in Shiras's files, and upon refreshing his memory, Shiras remembered that he had openly shared his inventions with the professor, provided a sketch, and wished him well in his endeavors.

Figure 2.5. German naturalist-explorer Carl Georg Schillings relied on Shiras's trigger mechanism to photograph African big game at the turn of the twentieth century.
Courtesy of Stadtmuseum Düren, Germany

Armed with this updated information, Roosevelt further clarified the matter with the ambassador. There is a long list of acknowledgements in the first (1905) English edition of Schillings's *With Flash-Light and Rifle*, but George Shiras's name is missing. To the ambassador's credit, the error was somewhat rectified in the second edition: Schillings credited an anonymous "American sportsman" with furnishing information on the method of night photography he used.

Shiras continued to share his knowledge of camera trapping and patented his inventions in the United States and abroad, but he allowed their "free use to the public to prevent their commercial exploitation."[9] Word of his work spread, but he did not publish on the set camera until 1913, when his article on the topic appeared in *National Geographic Magazine*.

A FRUITFUL PARTNERSHIP

Shiras had a large circle of friends who accompanied him into the field, and Frank Chapman was one who took up the practice of the set camera. Chapman, a banker turned ornithologist, was curator of birds at the American Museum of Natural History in New York, and the camera was a tool of his trade (Figure 2.6). By 1900 he had published *Bird Studies with a Camera*,[10] and the same year he and Shiras began exploring the use of the motion picture camera. Shiras found the negatives too small for satisfactory enlargements and lost interest in moviemaking, but the two continued to pursue photography together during field trips to Florida.[11]

In 1926 Chapman moved to Barro Colorado Island in the recently flooded Canal Zone, where he decided to survey the mammals.[12] The six-square-mile island was a result of the American enterprise of building the Panama Canal. The ensuing flood inundated 165 square miles of lowland forest and created several islands in a vast lake. To American biologists, the island was a gift for the study of tropical biology and was entrusted to the Institute for Research in Tropical America, a consortium of museums and colleges.[13]

Chapman made use of two flashlight mechanisms that wildlife photographer William Nesbit had left on the island and succeeded in recording agoutis, coatis, puma, tapir, ocelot, and collared and white-lipped peccary (Figure 2.7). Shiras praised the ornithologist's survey, noting that his account "should inspire others to make similar records of the rarely seen beasts that haunt the jungles of the American tropics."[14]

Figure 2.6. Ornithologist Frank Chapman dabbled in large-format camera trapping on Barro Colorado Island in the Canal Zone (now Panama). *Courtesy of the American Museum of Natural History Library*

WILDLIFE PHOTOGRAPHY ACROSS THE POND

With the exception of Schillings, few Europeans pursued flashlight photography with the enthusiasm of Americans. The reasons are not clear. Europe lacked the diversity of big game found in North America, but it certainly had its share of large mammals. Hunting in Europe was mainly an activity of the privileged class, while all classes hunted in the United States. Birding, however, was a popular pastime in England, offering ample subject matter for observation and photography, and in the 1890s several Britons began making their names as wildlife photographers.

One of them was Oliver Pike (1877–1963) (Figure 2.8). His photographic journey began as an understudy of Reginald Lodge, an ornithologist who made his living as a photographer. Lodge and Pike "experimented with electrical remote shutter releases and trip-wire releases, which brought them some success but were not the complete answer" to photographing birds up close.[15] Pike became a keen student of avian natural history and had a flair for filming that revealed his engaging subjects as less than harmless. His still pictures and movies won him considerable acclaim, and his knack for overcoming the deficiencies of the equipment led to the invention of the Birdland camera, manufactured by the London firm of Sanders & Crowhurst. His busy career was a major source of inspiration to amateur ornithologists and included twenty-five books, over fifty films, and countless lectures.[16]

Pike shared the genre of wildlife photography with fellow Britons Percy Smith, F. Martin Duncan, Richard and Cherry Kearton, and J. C. Bee-Mason; motion pictures were their modus operandi. Their collective work set the stage for authoritative documentation of nature that gave way decades later to the BBC's *Planet Earth* series.[17] For most of these photographers, set cameras were only one of many trade techniques. Far away in British India, however, one Englishman's enchantment with camera trapping inspired a valuable record that gives insight into the thrills and frustrations of the sport.

A

B

Figure 2.7. A, Chapman's trip wire sets for mammals were unprotected and prone to accidents. ***B***, But they yielded delightful portraits of white-lipped peccaries and other wildlife. *Courtesy of the American Museum of Natural History Library*

Ann Primrose Pike, Oliver Gregory Pike

A

Figure 2.8*A*. British photographer Oliver Pike on a photographic outing with his wife and lifelong assistant, Primrose. *Courtesy of the Oliver P. Pike Estate*

B

Figure 2.8*B*. Two pet badgers featured in Pike's film *Two Little Orphans*. His stills and movies sparked public interest in nature and birdlife in the 1890s. *Courtesy of the Oliver P. Pike Estate*

A CHAMPION IN THE FORESTS OF INDIA

When Frederick Walter Champion went to India at the age of twenty, he had already made up his mind to photograph tigers. Police work would seem an unlikely occupation for a nature lover, but in colonial India amateur naturalists came from all walks of life. Champion had discovered wildlife photography in the bird photos of Richard and Cherry Kearton, and the photos of African mammals by Schillings awakened him to the possibilities of camera trapping. The seed of his camera trapping idea germinated in the forests of northern India.

The Great War postponed his plans. He enlisted in the Indian Army, and nearly a decade passed before he had the opportunity to pursue his ambition as deputy conservator of forests in the Imperial Forest Service. It was a fortuitous career change. He was headquartered in Lansdowne, a small but popular hill station in the Garhwal Himalaya, or Siwalik Range, in what is now the north Indian state of Uttarakhand.

The Imperial Forest Service offered ample opportunities to shoot game and observe wildlife, but the battlefield had changed Champion's attitude. He readily conceded that hunting endowed wildlife photographers with useful qualities, but wrote:

> Those of us who went through the Great War saw quite enough of Death and all its horrors for one lifetime; and surely, oh! Surely, the innumerable difficulties that beset the animal-photographer at every turn should act as a spur to men who will never admit fatigue or defeat when hunting to kill.[18]

Champion believed that "continued success in photographing wild animals will be obtained only by those equipped with good apparatus, which they have thoroughly mastered."[19] Though the Brownie box camera made its American debut in 1901 and the 35mm camera was invented in Germany in 1914, Champion, like other camera trappers of the period, clung to the older and larger models. His night camera was a Sanderson Regular quarter plate camera, a folding bellows camera that used quarter plates (3¼ × 4¼ inches). The film holder accommodated two plates on opposite sides, and the cover slide had to be removed before exposing the plate.

No practitioner of flashlight photography found the process user-friendly. Champion wisely acknowledged that "if flashlight photography were too simple, it would soon cease to appeal, since it is one of the hard facts of life that a thing which comes too easily soon tends to lose its fascination."[20] To reduce "the heavy shadows which mar so many flashlight pictures," he used two to three flashlights purchased from William Nesbit.[21] He placed them according to standard practice, immediately behind and on both sides of the camera, 12–20 feet from the subject. They were battery-powered and wired to the shutter release. The six-inch focal length lens was set at full aperture with a **shutter speed** of 1/100th to 1/200th of a second. The results were pleasing crisp images with a diffuse lighting. Lacking are the sharp black shadows of the modern trail camera using only the camera's electronic flash.

Champion shunned the image distortion that photographers commonly seek today and went to considerable lengths to correct it "by enlarging at an angle, the enlarging screen being tilted so that the side of the negative which is too large becomes smaller in the enlargement than the other side."[22] He developed his negatives in the field and sent them for enlargement "should the fragile glass plate happen subsequently to get broke."[23]

Champion started photographing wildlife in the 1920s and continued until at least the mid-1930s. The flashlight camera was his primary medium of discovery, and his evocative images of Indian wildlife are the centerpieces of two books, *With a Camera in Tiger-Land* and *The Jungle in Sunlight and Shadow*. In both, he interweaved his observations of wildlife with his views of nature, hunting, taxonomists, captivity, and the ethics of flashlight photography. His views were decidedly at odds with prevailing utilitarian and anthropomorphic sentiments, but the books became classics of Anglo-Indian natural history.[24]

In photographing tigers in daylight, Champion found that for all his efforts he received very few acceptable photos. Dappled light and shade detracted from the great cat's magnificence, while images from flashlight sets stood the cats in bold relief. The set camera's notorious limitation was that it could take only one photo a night, and the chance encounter of tiger and camera was far less probable than of deer, wild pig, or porcupine. But he disdained the use of multiple cameras to enhance his chances. He believed in a more intimate relationship between the photographer and one or two cameras, rather than a dozen: "The whole art and interest of flashlight work lies in personally controlling one or two cameras and trying to anticipate what some particular animal will do."[25]

The logical solution was to locate a tiger kill or stake out a live buffalo as bait. The camera was surreptitiously set, with the remains in clear view. The window of opportunity was brief. A tiger can devour large prey in a couple days, and the smell of the carcass is a meal call to many other smaller predators, not to mention pigs. Champion set his cameras late in the afternoon to reduce the chance of crows, vultures, and other animals triggering the device. He wrote:

> The choice of site and adjustment of this trip wire are by far the most important points in automatic flashlight work, as upon them depend the success or failure of one's efforts. Almost every kill needs to be treated in a different way: sometimes it may be completely surrounded by the trip wire, sometimes one can be fairly certain of the tiger's probable line of approach, and sometimes the risk from birds,

> jackals and other creatures is so great that it is better to erect the apparatus in some other place, where the tiger is likely to pass during the hours he spends near the kill. [26]

Figure 2.9. Chicago lawyer Arthur Tappan Gregory relied on large-format cameras to document wildlife in the Great Lakes and beyond. He was the last of the three pioneering camera trappers to chronicle his adventures in popular books.
Courtesy of the Denver Museum of Nature and Science, IA.501.BAILEY.31

Then he camouflaged the camera and "by judicious arrangements of thorns and branches"[27] created an avenue to the trip wire. He was baffled by the number of times the tigers abandoned their kills, until he realized that many sets failed because the tiger had seen him at work. He obtained thrilling images of many species, and even today his black-and-white images of tigers are as evocative as the best of contemporary tiger photos.

Champion confessed that "to the keen student of animals," sitting over a tiger kill offered more insights and observations than camera trapping, but the downsides included staying awake in the wee hours, trying to see anything on moonless nights, and sitting still while being eaten by mosquitoes, with the long-term detrimental effects of malaria. Using a set camera bypassed these detractions, but learning from the experience was more like a cat-and-mouse game.

Perhaps more than any other camera trapper of the period, Champion had a scientific approach. Photographing tigers without resort to baits and natural kills, he wrote, "requires an intimate knowledge of the locality, as well as of the habits of the particular individuals one wishes to photograph."[28] He studied the daily rounds of tigers to increase his chances of tiger–camera encounters: "In order to obtain information on this subject, I have had maintained daily records of tracks passing definite fixed spots for six months or more and have thereby added to my knowledge of the movements of tigers and other denizens of the jungle."[29] After about 150 attempts to photograph tigers on their nocturnal beat, he found they passed the camera only six times, yielding only four photographs, but he acknowledged that one was the best he ever took.

Experience informs the camera trapper, and Champion learned that leopards were more challenging photographic subjects than tigers. They were not good subjects on trails and roads because they had an uncanny ability to detect "any but the darkest and thinnest wire."[30] Unlike their village counterparts, jungle leopards were not drawn to live goats used as bait. When tigers killed goats in the jungle, they polished off the meager fare in a single night, so even hungry leopards were denied a free meal.

A CHICAGO LAWYER'S HOBBY

A Chicago lawyer would seem an unlikely devotee of camera trapping, but Arthur Tappan Gregory made up for lost time when he took up wildlife photography in his midthirties (Figure 2.9). Like

Figure 2.10. Robert Sturgis, Gregory's loyal friend and field assistant, was the gifted tinkerer who built and repaired Gregory's camera triggers and field darkroom kit.
Courtesy of the Smithsonian Institution Archives

Shiras, Gregory came from a privileged family and shared with his brother Stephen a keen childhood interest in natural history. His father sent him to prep school, and after graduating from Yale, Gregory got his law degree from Northwestern University and began his law practice in the Windy City. When the United States declared war on Germany in 1917, Gregory enlisted in the army and shipped off to France as a first lieutenant in the 313th Artillery. He returned from France when the armistice went into effect two years later and resumed his family practice at the age of thirty-four.[31]

By then over two dozen popular books had been published featuring photos of wildlife, and it is likely they stimulated his interest. He began camera trapping on fishing trips in the greater Chicago region with his brother and his friend Robert L. Sturgis (Figure 2.10). Stephen Gregory was an avid birder, and Sturgis was a gifted tinkerer and devoted friend who readily tackled various tasks that "Tap" requested. His most frequent assignments were to construct the flashlight triggers originally designed by Shiras.

Gregory joined the American Society of Mammalogists. Scientific journals were less technical in those days, and editors indulged a certain amount of poetic license from their contributors. Gregory's article in the society's journal, "Flashlights of Small Mammals," featured photos of mink, beaver, and striped skunk taken near Lake Superior, described the photographic apparatus, and captured the adventure of the experience, including the terror-stricken flight of a domestic cat. Like Frederick Champion in far India, Gregory's writings tell the trials and tribulations of an oft-frustrated devotee of flashlight photography.

Over the next seven years he wrote seven articles for the journal, featuring photos of deer, rabbits, and rodents, from white-footed mice to beavers. Then he expanded his quest for new species and moved from camera-friendly rodents and lagomorphs to elusive carnivores. The American mink proved a formidable quarry. "Always fresh mink tracks induced the making of the set,"[32] he wrote, and he used all the trapper's tricks along northern Michigan's creeks and lakes. He often prebaited a site before setting the camera and used an "infinite variety of bait," from various species of fish and fish guts to cod liver oil, crayfish, and mussels. He tried carcasses and body parts of songbirds and rodents, and even butchered meat. He sweetened the pot with irresistible attractants, such as beaver castor and musk and the old trapper's standby of fetid scent—a gagging decoction of rotting meat and scent glands.

Over a period of fifteen years, Gregory and his friends vacationed on the shores of Lake Superior and searched for mink. They found their tracks in the mud, and they saw them in the flesh, but only one set rewarded him with pictures. The best of the four photos, he admitted, was "not a perfect picture by any means," but it revealed the crafty predator "sneaking up on the bait."[33] Camera trappers like Shiras had often succeeded in getting mink, but Gregory was at a loss to understand the reason for his own failure.

Such frustrations were not defeating because getting the perfect picture was not this camera trapper's sole purpose. The chase was as important as the kill, and even the mundane white-tailed deer or raccoon could save the day by presenting a suitable image. In due course Gregory set his sights on more elusive species.

Figure 2.11. Twenty-eight-year-old Stanley P. Young returns to camp with a trussed ninety-eight-pound Mexican wolf in 1917. As chief of predator control for the Biological Survey, Young was instrumental in launching expeditions for Gregory to camera trap wolves and puma in the late 1930s.
Courtesy of the Smithsonian Institution Archive

LAWYER MEETS TRAPPER

In 1929 Gregory met Stanley P. Young. It was a chance encounter during a fishing trip to the Madison River in western Montana and a pivotal event in the lives of both men. Young, in charge of the Biological Survey's Division of Predatory Animal and Rodent Control, was on rounds with his staff. He was three years younger than the forty-four-year-old lawyer, and not a fisherman, but they struck it off well. Gregory showed him his set cameras and explained his camerawork. He later wrote that Young's "lively imagination and keen perception showed him the broader range of possibilities in this method of hunting and his interest seemed at once aroused."[34]

Coyotes were Gregory's target species on this particular trip, and predators were Young's specialty. Educated as a mining engineer with a master's degree in biology, Young had forsaken a teaching career to hunt predators (Figure 2.11). His homespun demeanor endeared him to the cattle and sheep men of the western ranges. His superiors recognized his leadership, and he rose through the ranks of the Biological Survey. And he could tell a few stories, like the time Pancho Villa's men captured and held him hostage while he was tracking a wolf dragline in Sonora, Mexico.

Gregory knew that Young could open the doors to new camera trapping adventures and made known his interest in timber wolves. But finding a wolf population was not easy. By the mid-1930s US predator control agents had exterminated most populations. Wolves were getting scarce on the cattle ranges, and contract trappers were in constant pursuit of Colorado's last loner wolves.

Young enlisted A. E. Gray, district agent for Oklahoma, Missouri, Arkansas, and northern Louisiana, to locate a wolf population, and Gray put his trappers to work. One of them found a pack still living peacefully on the 82,000-acre Singer Wildlife Refuge, bisected by the Tensas River in Madison Parish, Louisiana.

Four years and several encounters later, Young wrote to Gregory that it was time to take action and set a date. They allowed themselves sixteen months to prepare, and letters went back and forth between Chicago and Washington, DC. Gregory expanded his inventory of cameras to ten and recruited Robert Sturgis to design and assemble triggers and to build a portable darkroom kit.

Young educated Sturgis on the trapper's fine art of hiding steel traps on predator runs, and Sturgis designed a pressure plate trigger that could be buried on the trail.

The expedition began on October 6, 1934, when Gregory and Frances Dickinson, president of the Chicago Academy of Sciences, rendezvoused with Young and his team of state and federal staffers at the camp. With newly made benches and tables, the camp was shipshape, and Tom the cook was already preparing the noonday meal. Eager to start, Gregory held a seminar on the virtues and vices of the set camera, "and how best to adapt it for field use on wolves."[35] After making practice sets, they turned in to a chorus of barred owls.

They set to work the next morning, and by day's end had found thirteen wolf scrapes and had set six cameras. Gregory selected a suitable tree 12–18 feet from the scrape and set up the camera. Young and the men dug the trench for the wires and treadle while Sturgis wired the trigger. The trappers wore cotton gloves, kneeled on a manure-soaked cloth to mask their scent, and scattered the excess soil 150–300 feet away. Gregory disguised the camera, and the final touch was the judicious application of wolf urine to capture the interest of the carnivore trotting on its nocturnal rounds. The finished site looked untouched.

At eight thirty that evening they were rewarded when "the boom of a distant flash came to our ears." The next morning they found that the set had been hurriedly made, but Gregory knew that sets "hastily thrown out as an afterthought, may bring the finest results."[36] The developed image showed a melanistic, or black, wolf that became the expedition's icon (Figure 2.12). Gregory titled his account "The black wolf of the Tensas" and wrote: "We had accomplished our mission within thirty-six hours after reaching camp and within six hours from the completion of the set that was sprung."[37] This was likely the first camera trap photo of a wild wolf, and it was the only photo taken at that set.

The team worked daily during the remaining three weeks on the camera trap line, and their reward was thirty-six photos, four of which were of black wolves. Nights on the Tensas River no longer ring with the calls of black wolves, and their status as a species has been debated. But the image of the trotting black wolf, nose to ground, reminds us of a bygone era when wolves roamed the Louisiana bayous.

Figure 2.12. *The black wolf of the Tensas*, one of several photos Gregory shuttered on trails in the Singer Wildlife Refuge, Louisiana, in 1934. *Courtesy of the Smithsonian Institution Archives*

LIONS OF THE SIERRA DEL CARMEN

Figure 2.13. Campsite in the Sierra del Carmen, Coahuila, Mexico, where Gregory camera trapped mountain lions.
Courtesy of the Chicago Academy of Sciences

In less than three years, another camera trapping expedition was underway to the Carmen Mountains in the state of Coahuila, Mexico. Photographing mountain lions was Gregory's idea, but Young picked the site and arranged for provisions and permissions from the Mexican government. He billed it as a joint venture between the US Biological Survey, the Chicago Academy of Sciences, the Smithsonian Institution, and the National Zoological Park. The Department of Agriculture announced the objectives in a press release: wildlife photography, collection of mammal specimens, and capture of live mammals for the National Zoo.

The Sierra del Carmen is a **sky island**, an isolated 40-mile section of the Sierra Madre Oriental cloaked with oaks and pines and surrounded by the Chihuahuan Desert. Zoological collecting was justified because two intrepid naturalists, Dr. E. W. Nelson and Captain E. A. Goldman, had failed to visit the area during a monumental eight-year collecting expedition in the 1890s. There was also a practical reason—wealthy Texans of the Carmen Mountain Hunting Club had offered use of their clubhouse and a hundred thousand acres in the Sierra.

Gregory's account of the expedition is a bedtime story for camera trappers. It paints an evocative picture of expectations, hard work, and the trials and tribulations of using mechanical triggers and keeping the flash powder dry.

The party was eight strong. In addition to his assistant Sturgis, Gregory invited a prominent Chicago surgeon named William Hazlett. Young assigned three hardy staff from the US Biological Survey, C. R. Landon, Antonio Gonzales, and lion hunter Ira Wood, who was also an acclaimed goat-roper. He sent the colorful hunter to reconnoiter the Sierra del Carmen three weeks before departure.

With ten cameras, gear, and baggage, Gregory and Sturgis embarked on the four-day car trip from Chicago to San Antonio in early September 1935. Young arrived by train, and after transferring the gear to a truck and two survey cars, they crossed the border "with courteous and efficient treatment."[38] After a night in Villa Arcuna, they negotiated 127 miles of dirt roads crossed by arroyos, mud holes, and a small lake with a flock of pintail ducks. They arrived before dark at the clubhouse, at an elevation of 5,000 feet, and the next day sorted gear and baggage while the club's retainers rounded up and shoed the thirteen burros and one mule that carried them into the mountains the following morning (Figure 2.13).

Wood located **pugmarks** and scrapes, but cold rains interrupted the timely deployment of all the cameras. The sets were similar to those used for wolves, except for the use of catnip oil as lure. Under Young's authority, the Biological Survey had prevailed on the Bureau of Plant Industry to grow sufficient catnip to test extracts as bobcat and mountain lion lure. Field tests proved that the potion still attracted pumas six months after application.[39]

The ensuing days were a rollercoaster ride of hope and disappointments, as foggy cold rains alternated with halcyon days and majestic views of the Chisos Mountains beyond the Rio Grande. Small teams rode the trap lines daily and "rushed [the exposed plates] to the portable darkroom, one of the results of Sturgis' inventive genius."[40] Rabbits, deer, and bobcats triggered only a few cameras. Gregory tried to understand the bad luck:

> It was hard to believe that no lions were revisiting any of those numerous fresh scratches which had so heartened us at the outset. We must have come into the mountains at the very inception of a great exodus to lower elevations. It seemed that every lion upon our advent had found that his schedule called for another tour of his extensive rounds.[41]

They abandoned inactive sets, made new ones, and Gregory voiced a camera trapper's lament: "Too often we are reluctant to move one already out. Once a camera is on duty, it always seems that surely the next night some desirable subject will fire the flash."[42]

Nightly in the starlight, they listened to the whippoorwills and wondered what the next day would offer. More rabbits, deer, and bobcats were in the offing, but fortuitously a lion killed a goat on a nearby ranch. They arrived a day late, and the meager remains yielded only a photo of a ringtail. Adding insult to injury, a mounted vaquero dragging a pole tripped two other cameras "with uncanny and unerring precision."[43]

With three days left, Gregory mused that "it was not easy to be philosophical." The chances were slim that a lion would pass any remaining camera, and the departure of the Biological Survey's Hazlett and Landon was a "sobering incident to those of us who remained."[44]

Finally they got a blurry image of a large tom swinging his head and lashing his tail. "We could not be sure whether he was rushing through or trying to skid to a stop for a whiff of catnip."[45] Young measured the cat's leap upon experiencing the explosive flash: 13 feet. Perhaps the tide was turning. Another plate magically revealed the tail end of another large cat. That camera had set for sixteen nights before the cat arrived. On October 6, lion trapper Wood and Mariano delivered another camera's plate to camp. Gregory developed the negative and found the last image of a lion. All's well that ends well. But Gregory wrote:

> Depression at the idea of leaving stole into our thoughts. The skies joined our mood and by mid afternoon, wept upon us disconsolately, repeating this manifestation on and off during the night, leaving the valley as it appeared to us the next day, filled with fog under heavy, low-hanging clouds. Into such a prospect we drove soon after noon on the tenth, saying goodbye reluctantly to our loyal packers, our study and hardworking horses and the surprising little burros.[46]

The expedition was the first time that camera traps were used in a professional survey of wildlife, in this case together with scientific collecting. Chapman had already laid claim as the first person to camera trap a mountain lion. Gregory might have hoped for better photos, but the trip was a success. The men deployed ten cameras across 15 miles of trails as the crow flies, almost always at lion scratches. If a fired camera was left unchecked, it was out of operation. So they checked them almost daily from September 15 until October 8 and refreshed each set with catnip as often as possible, often every day when it rained. When rabbits or inactivity prevailed at chosen sites, they moved the cameras to new locations.

Gregory's account of the trip in the *Chicago Naturalist* contains sufficient detail to estimate that the ten cameras were active about 194 camera trap days. During that time they obtained three pictures of mountain lions, two of bobcats, two of ringtail, three of white-tailed deer, and an unknown number of rabbit portraits, not to mention the errant cowboy and goat.

Gregory did not mention that Bob Sturgis, in addition to his other duties, had managed to trap and shoot a dozen rodents and prepared them as specimens for the US National Museum. Other team members salvaged skulls of black bear, mountain lion, and bobcat, thus fulfilling the trip's objective to collect specimens. From these specimens Edward Goldman described three new rodents and named a new species of pocket gopher, *Thomomys sturgisi*.[47]

ATTITUDES AND ETHICS

The popularity of camera trapping in the first half of the twentieth century is hard to gauge, but there were more camera trappers than the few who immortalized the pursuit in writing. Gregory hinted at its popularity in the Great Lakes: "During those never-to-be-forgotten nights, we saw from time to time, throughout the hours of darkness, white flares of light flashing in the sky to the north, where other votaries of the sport of wild animal photography were bombarding the bears six miles from where we sat."[48] There were enough enthusiasts for an enterprising photographer named William Nesbit to build and sell flashlight triggers, lamps, and a moisture-proof camera.

But cameras were already expensive, and the added costs for flashlight cameras made camera trapping less accessible to the working class. It was a hobby for those with means, and its advocates promoted it as an alternative if not a substitute to hunting. The pioneers often endorsed their sport as a form of hunting that demanded no bloodshed. In Champion's words, "Flashlight or any other form of wild-life photography is a magnificent sport which calls for the maximum of skill and woodcraft, and, in some cases, pluck as well. In proper hands it must be infinitely less cruel than any kind of hunting."[49] The reward of the chase was often a painterly record of nature rather than a pack of cold meat.

Suffering is a critical measure in ethical debate, and its detection has always been a subjective call. Pioneers might proclaim camera trapping as largely nonlethal (baits excepted), but whether it caused suffering was another matter. In 1914 Welsh-born naturalist Arthur Radclyffe Dugmore wrote that beavers repair breaches in their dams with remarkable promptness. For three weeks he broke a dam as many as three times nightly and recorded the beavers' appearances with flashlight cameras.[50] Some might find his experiment a cruel and unnecessary disturbance, while others would defend it as harmless or enlightening.

Wildlife photographers had their own detractors. Champion noted that they were blamed for disturbing nesting birds and for "harrying the big game of Africa by trying to take photographs from low flying planes."[51] Hunters claimed that photographers frightened game and blamed them for declining success. Shiras confirmed some of these allegations when he wrote that the deer of Whitefish Lake learned to avoid approaching jack lanterns, which they evidently associated with the frightening boom of the flash. But he offered no apologies. He had witnessed the decline of white-tailed deer to market hunters who used jacklights.

The explosive flash was an ethical issue even among set-camera photographers. Chapman was never at ease with its disruptive force: "It is difficult to conceive of a more terrifying experience than this sudden encounter with a blinding glare of light and cannonlike report at a distance of 10 or 11 feet. This is the one objectionable feature of camera-hunting at night."[52] He was particularly troubled by Shiras's practice of setting a second flashlight camera to capture images of the animal's reaction to the explosion.

> To yield to our curiosity to see "what happens" is to destroy this feeling of seeing while unseen and, at the same time, to take unfair advantage of the animal. It is bad enough to frighten a creature; to photograph it in the act of responding to the uncontrollable impulse of fear is adding insult to injury.[53]

Shiras took exception: "This sympathy for the frightened animal I believe to be mistaken, for I doubt that the second explosion adds much to its consternation."[54] Champion condoned the use of flash as long as its effects weren't detrimental, but he drew the line at flashing at water holes during the dry season and at dens and nests containing young animals.

> In every case the photographer, if he be a true sportsman, must decide for himself whether any scheme he may evolve for obtaining a photograph is fair on the subject or not; and if his conscience tells him that it is not, the chance must be let pass no matter how tantalizing it might be.[55]

Champion dwelled on nature and ethical issues in the last chapter of his second book. Is nature a place of death and terror or of a different essence altogether? He discussed the senses of the tiger, speculating about its subjective world, but he seemed to concede to the inscrutable and fell back on the words of poets.

THE AGING PIONEERS

The pioneers of camera trapping were in their autumnal years when the looming clouds of World War II darkened the mid-1930s. Nesbit and Schillings had already faded from the scene. Shiras consolidated his many writings into a two-volume work in 1935. *Hunting Wildlife with Camera and*

Flashlight was a window into his life and an environmental history spanning sixty-five years. Nine years later he died in Marquette, Michigan, at the age of eighty-four.

By 1938 Chapman had long since given up camera trapping when he published his last book, *Life in an Air Castle: Nature Studies in the Tropics*. This recollection of natural history observations included his story of camera trapping in Panama. He retired from the American Museum at the age of seventy-nine, three years before his death in 1945.

Gregory, who turned fifty in the 1930s, was immersed in civic affairs, but still found time to pursue his hobby, write articles, and publish two books on camera trapping. During the war, he was elected president of three bar associations: the Chicago, Illinois State, and American. When the war ended, he served as an observer and photographer at the Nuremberg trials. His last book, actually a pamphlet, *Nature Photography at Night*, was published in 1957, four years before he died.

Champion, the youngest of these men, moved to East Africa after Indian independence in 1947, and there he continued his career as a forester. He retired to Scotland six years later. At an Indian forestry conference in 1961, he confessed that both he and his wife were at times "overcome by an overwhelming feeling of nostalgia for the glorious foothill forests of the Himalayas."[56] He continued to write until 1969, when he died at the age of seventy-seven.

CHAPTER 3

Camera Traps for the Masses: 1946–Present

As the pioneers of camera trapping faded from the scene in the years after WWII, wildlife biologists in the United States and Canada started converting new camera models for remote use in ecological research. These camera traps, despite limitations, were excellent research tools—but their construction was intimidating, which limited their use. Building repeating automatic camera traps became easier in the 1980s, when electronic **point-and-shoot** cameras and **infrared** sensors became available.

By the end of the decade, the trail camera industry emerged as entrepreneurship's answer to North America's hunters. The commercial trail camera made simple camera traps accessible to field biologists and outdoor recreationists in general, marking the dawn of a diverse second generation of camera trap enthusiasts. Some were veterans of the war, others were younger, but most were wildlife biologists who were familiar with the writings of the pioneering camera trappers. They began experimenting with modern cameras for remote studies of animal ecology.

Oliver "Paynie" Pearson belonged to this second generation of camera trappers (Figure 3.1). Widely admired as a scientist at the University of California in Berkeley, he had a gift for investigating problematic issues, from the taxonomy of South American rodents, to the energetics of lactation in shrews. His traffic survey of field mouse runways in the 1950s was the first technical study of wildlife ecology based on camera traps.[1]

Game trails have always interested naturalists because they tell us where wild animals move and how they use their habitat. Trails come in all sizes, are usually engineered for energetic

Figure 3.1. In the mid-1950s, Oliver Pearson initiated camera trap investigations of vole runways in his backyard in Orinda, California. This 1960s portrait shows the professor during field studies in South America.
Courtesy of the O. P. Pearson family

economy, and are most pronounced where activity is concentrated—near feeding areas, water, or mineral licks. Climb a tree or hide near a game trail, and sooner or later you will see who uses it.

Pearson wanted to know what goes on in the microcosm of meadow mouse, or vole, runways. To a mouse, a patch of weeds or a grassy field is a vast jungle, and species like meadow mice bushwhack and maintain their own runways using their teeth. In an open runway, they zip along at a brisk clip and tend to their daily business. Freeloaders like harvest mice and shrews profit from the labors of their larger neighbors.

Pearson's property in Orinda, California, harbored populations of meadow voles and harvest mice only 50 feet from the house, and the epicenter of his study was a 20 x 20-foot weed patch with a brush pile in the middle. His trail cameras were two 16mm movie cameras modified for single frame exposures in tandem with weather stations to survey the traffic. Two trigger devices, a treadle, and a photoelectric cell activated the cameras' shutters with equal success whenever a mouse or even a large insect passed the camera lens. The cameras were synchronized with electronic flash units, but to avoid frightening the mice, he masked the flash with eighteen layers of red cellophane. He noted that "a muffled clunk made by the mechanism seemed not to alarm the mice unduly."[2]

Pearson positioned a camera unit on one side of a mouse runway. On the other side, and framed within the camera's view, was a miniature weather station—a dial thermometer, a hygrometer, a ruler, and an electric clock with a sweep second hand (Figure 3.2). Whenever a creature took its own portrait, it also recorded its body length, the time, and a weather report. Since he live-trapped the mice and gave them distinctive "haircuts," he could identify many of them individually.

Powered with alternating current, the cameras ran day and night, but there was a problem with differential exposure. Daytime pictures were overexposed because the shutter was slower than 1/30th of a second and the **f-stop** was adjusted for night. Red cellophane was again the solution—this time over the camera's lens. The units were sheltered in glass-fronted housing with overhanging eaves and a blackened light bulb to counteract the effects of rain and dew.

Figure 3.2. A frame of a harvest mouse from Pearson's 16mm camera trap at the family's property.
Courtesy of the O. P. Pearson family

Not long after Pearson deployed the equipment, the neighbor's Siamese cat discovered that the camera housings were superb perches for mousing. Pearson solved the problem by fencing the weed patch. Then a slender salamander electrocuted itself while short-circuiting the treadle. Delusional song sparrows were another problem. They wasted a lot of film shadowboxing their own reflections on the weather station window. Curiously, one bird sidestepped an oncoming meadow mouse, a gesture of road etiquette rarely seen on California highways.

In nineteen months of operation, the camera traps took 8,495 photographs. In other words, wildlife made an average of eleven passages per day through the runways. The professor noted: "A patient, non-selective predator waiting for a single catch at runways such as these could expect, theoretically, a reward each 2.2 hours."[3] The cameras mined enough information for three scholarly articles.[4]

Pearson's pleasant surprise was that twenty-six species of animals, including weasels, snakes, and mole crickets, made use of the runways. He never saw three of these species within at least a mile of the study area, proving that camera traps often disclose nature's little secrets. Ironically, he never photographed three other species that he commonly saw within 100 feet of the recorders, proving also that camera trap location is critical to success.

Though two species of mice used the runways, Pearson's cameras revealed that meadow mice were the public works custodians of the small mammal community. When meadow mice stopped using a runway, weeds and seedlings quickly filled it in. Another curious finding was that mice apparently used runways selectively, even when nearby runways seemed equally suitable.

Pearson's paper stimulated other biologists to develop similar cameras, and, in all cases, automatic operation was the primary objective.[5] Invention and tinkering were requisite skills. Academics and graduate students experimented and developed their own automatic cameras, and some even published the results, but how many people used the published plans is unknown. Terminology and circuit diagrams were daunting to at least some who dearly wanted an automatic camera for wildlife photography and research.

NEED IS THE MOTHER OF INVENTION: CAMERA TRAPS FOR RESEARCH

Like Pearson, other biologists with the requisite know-how built their own camera traps and added features to suit their research needs. A University of Maryland doctoral student named Nicholas Smythe, for example, developed a camera trap in the late 1960s before embarking on a two-year ecological study of the agouti, a rabbit-sized Central American rodent. To determine its activity cycles, he needed an automatic camera that could take a large number of exposures.

> I tried several different cameras before settling on the Olympus PEN camera. The details are foggy in my mind, but I believe that the first was the Canon Dial. Then there was a Russian-made camera, which proved unsatisfactory for reasons unremembered. I think there were others too, but I settled on the split frame 35mm Olympus PEN camera, which was one of the first motorized 35mm cameras that advanced the film automatically.[6]

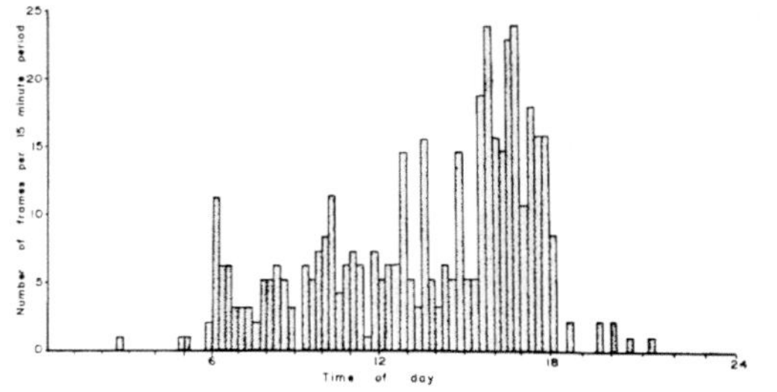

Figure 3.3 Camera trap data of agoutis clarified their home ranges and activity cycles. The analyses were published in Smythe's doctoral thesis.

Smythe knew that his subjects and other fruit-eating species, such as spiny rats, **pacas**, coatis, and peccaries, would quickly expose a roll of film. Film cameras were good for twenty or thirty-six exposures, but half-frame cameras doubled the maximum number of exposures. In fact by bulk-loading his own film canisters, he could push the half-frame cameras to take upward of seventy-two frames per roll.

The triggers were a constant problem. He wrote:

> I made some that used a 3–4mm copper ring as one terminal and a vertically hung copper wire as the other, but they were unreliable. So I used trip wires and micro-switches, and they were moderately successful, but peccaries and tapirs did not treat them kindly. And I have had several pictures that showed, indisputably, that agoutis knew very well where the wires were, and inadvertently tripped the cameras while attempting to avoid the trigger wire. I did rely on photoelectric sensors. These, of course, required a light source on one side and a sensor on the other, rather than a light and a return reflector. I didn't know if IR [infrared] light sources were available at the time, so I used **incandescent** (modified flashlights) for the light. I made them myself and they suffered accordingly (the atmosphere in the Panamanian forest is not kind; I still have ear and body fungi that are not diagnosable by local MDs).[7]

Smythe **hacked** four cameras, added electronic flashes, used **solenoids** to release the shutters, and housed them in water-resistant boxes. Studying activity rhythms requires knowing when photos were taken, of course, but time stamps and **EXIF** files didn't exist at the time. "I was not able to find a 24-hour DC clock, so I used one that relied on AC. This involved running long wires from the laboratory to the camera sites which, in the ambient 98–100 percent humidity, gave rise to some interesting involuntary muscular contractions."[8]

The negatives contained hundreds of pictures of peccaries, agoutis, pacas, and coatis blithely foraging in the **windfall** of fruit from giant tropical trees, not to mention ocelots, motmots, iguanas, and a boa constrictor. The cameras also reaped the data Smythe wanted on activity cycles and revealed the agouti's role in dispersing seeds of tropical trees (Figure 3.3). And they made it possible to estimate the size of the agoutis' home ranges by recording marked animals in different areas. Smythe's study was certainly one of the first PhD dissertations made possible, at least in part, by camera traps.[9]

By the mid-1970s a variety of 35mm and medium-format cameras offered more options to researchers in need of camera traps. One of those researchers was Paul Joslin, a young Canadian who had just completed PhD studies of wild lions in India (Figure 3.4). India's Gir Forest National Park harbors the only known remnant of a lion population that once occurred from Africa to South Asia, but Joslin wondered if lions still existed in other remote corners of Southwest Asia, like mysterious Iran. To find out, he paid the country a visit, met with officials, and learned that the Wildlife Research Division just happened to be looking for someone to direct a new carnivore survey unit. They offered him the job. Joslin hurried home, finished his degree, and returned to start work before the ink had dried on his dissertation.[10]

The authorities wanted to know if Asiatic lions and Caspian tigers still existed in their country. The last record of a lion was an animal shot in central Iran in 1942, and the last record of a Caspian tiger was an animal shot in the 1950s. The prospects were not good, but Joslin was hopeful because "the Alborz Mountains where the tiger last existed was still well forested, had remote areas, and still supported a large population of wild boar."[11]

Figure 3.4. In the mid-1970s, Paul Joslin used mechanical bait triggers and Polaroid Land Cameras to survey Caspian tigers and other carnivores in Iran.
Courtesy of Paul Joslin

Joslin trained dozens of the region's field officers to differentiate tiger from leopard tracks in snow, and they made numerous plaster casts of pugmarks for later verification. "In two areas the cat tracks were so extraordinarily large that they could be tiger," he wrote. "I was blown away by the findings, and wanted to confirm one way or other their presence or absence, and that's when I hit upon the idea of putting out cameras tied to wild boar carcasses."[12]

Figure 3.5. A. One of Joslin's cameras: Agfa Autostar x-126 with cube flash and baited trip wire.
Courtesy of Paul Joslin

Camera choices were limited. Those requiring batteries to operate the flash weren't suitable in the freezing nighttime temperatures. Joslin settled on a box camera that operated a flashcube by mechanical action and required no settings (Figure 3.5*A*). It was also the cheapest option.

Transporting wild boar carcasses to the remote sites proved to be a strenuous challenge, but Joslin recalled:

> We were rewarded with a leopard photo on the first night of testing, which really spurred on my belief that a photo trapline system could work. Two winters of baiting in the two most probable areas produced photos of leopard, wolves, brown bears, and a host of other animals, but no tiger (Figure 3.5*B–D*). I concluded that exceptionally large leopards made the oversized tracks while walking downhill, which spread their toes to the max. Melting enlarged the tracks before we made the casts. I realized the Caspian tiger was most probably extinct.[13]

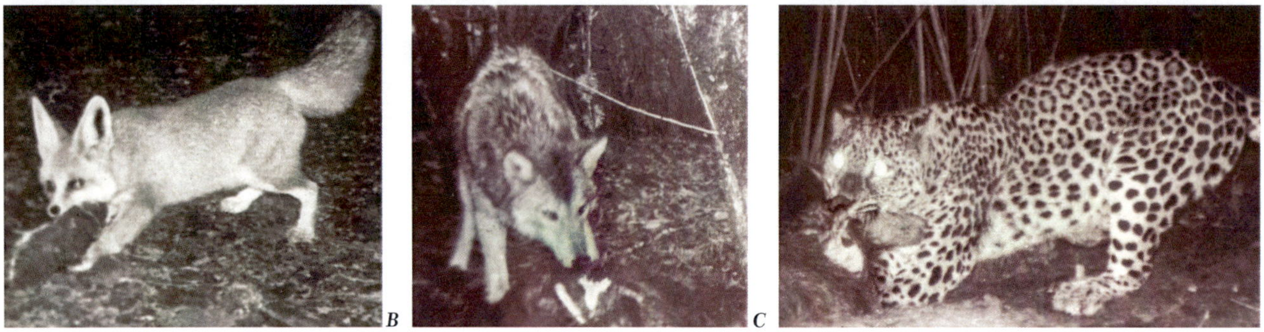

B C D

Figure 3.5. Three of Joslin's photo-captures: ***B***, Rüppell's fox *(Vulpes rueppellii)*; ***C***, wolf *(Canis lupus)*; ***D***, leopard *(Panthera pardus)*
Courtesy of Paul Joslin

As the surveys continued, with a focus on smaller carnivores, Joslin improved the technique. He experimented with the use of trip threads strung at varying heights across game trails, and he increased the mechanical efficiency of the triggers, which were an arrangement of strings and levers between the shutter release and the bait. He also added a weak point as a safety valve in the twenty-pound test line attached to the bait. This fixed the annoying problem of animals absconding with the camera attached to the bait.

The poor-man's cameras were not without headaches. They had unreliable shutter speeds and had to be advanced after each exposure. According to Joslin, "You couldn't see the result without taking the film out of the field to be developed." In fact he "usually pulled a roll for development after just a couple of photos had been taken, which meant going through a lot of film."[14] Eventually he switched to the cheapest models of **Polaroid** cameras that used mechanically operated flash bars. "I didn't have to worry about batteries running down quickly, because power was used only when processing photos. Not only did Polaroid provide instant results, I could take two or three photos per camera by using as many baits."[15]

The economy-class camera trap still did the job, and after three years, Joslin returned to Canada to pursue a more conventional zoological career.

In the 1980s zoology professor Donald Griffin decided to investigate the private lives of beavers (Figure 3.6). Griffin was on a quest for evidence of complex thought in animals, and though old enough to retire, he was busy challenging the conventional wisdom of comparative psychologists with the energy of a graduate student.[16] His plan was to investigate communication among family members in a beaver lodge and to test their reactions to a source of alarm.[17]

Griffin needed a sophisticated camera to meet the challenge, and with the help of a university engineer, he built a video camera with a source of infrared illumination and a remotely moveable head, which they deployed at Mink Pond at Harvard's Concord Field Station. The camera was remote, but it required an assistant to track the animals' movements within the lodge. Unfortunately, Griffin didn't publish details about the setup or the findings, but he made the footage available to Professor Dietland Müller-Schwarze of Syracuse University. Müller-Schwarze studied the film for a book coauthored with Lixing Sun titled *The Beaver: Natural History of a Wetlands Engineer*.[18] This was probably one of the most sophisticated cameras designed for a specific wildlife project, and it happened years before the British and Canadian Broadcasting Corporations (BBC and CBC) and amateur filmmakers had tackled the problem of recording behavior in the beaver lodge.

CHAMPION'S FOLLOWERS

While American academics were converting modern cameras into camera traps in the late 1970s, an American named Charles "Chuck" McDougal began photographing tigers with camera traps (Figure 3.7). An unpretentious ex-marine, McDougal did graduate research on cultural anthropology in India and Nepal, which answered his inborn wanderlust. He also hunted problem tigers in Nepal in the late 1960s. He returned to Nepal after receiving his PhD, and when the Convention on International Trade of Endangered Species (CITES) made tiger killing illegal in 1972, he started to study tigers. He was ideally situated to do so. He was then director of wildlife at **Tiger Tops**, a jungle lodge in Chitwan National Park which he comanaged with Jim Edwards.[19]

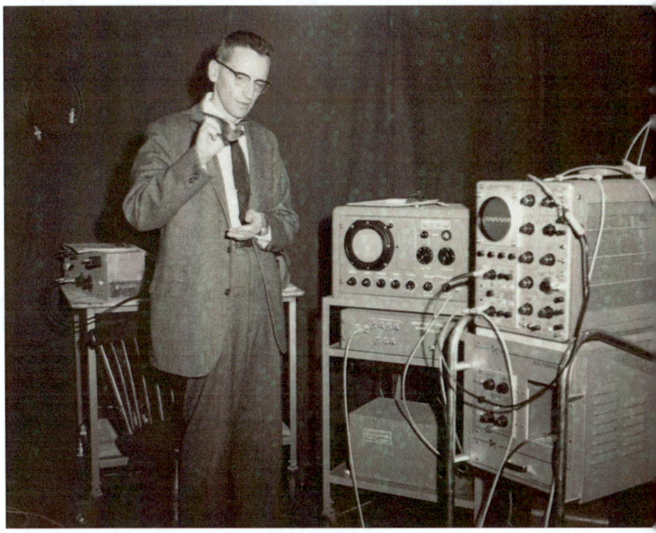

Figure 3.6. In the 1980s, Professor Donald Griffin used a remote camera to monitor beavers within their lodge at Harvard University's Mink Pond. In this photo from 1960, Griffin poses with equipment he and colleague Robert Galambos used to demonstrate echolocation in bats.
Courtesy of the Harvard University Archives, HUP Griffin, Donald R. (3), olvwork245838

McDougal was inspired by Champion's books and photos, but he didn't know exactly how to adapt the modern SLR camera as a camera trap. That changed when a crew from the British series *Anglia Survival* was filming tigers in the park. Cameraman Mike Price set a Nikon F2 on a game trail, wired the camera's remote jack to a pressure pad, and buried the pad and cable. When Price finished his assignment, McDougal started photo-capturing tigers with his own equipment and trained a small team of local trackers to monitor game trails for pugmarks and scent marks.[20] His trackers meticulously measured pugmarks to identify adult tigers, but specialists wanted to see validation of the method, and the burden of proof was on McDougal. He and his colleagues validated the method using camera traps. They tallied 8,683 trap nights of effort and identified over twenty different tigers.[21] Each year new tigers appeared, some fleetingly and never to be seen again, others to challenge and sometimes defeat incumbent territory-holders.

While McDougal photo-captured Nepalese tigers in the 1970s, a young Indian named Ullas Karanth was considering a change in career. Disillusioned as an engineer, he wanted to improve tiger survey methods and contribute to their conservation (Figure 3.8A). He had always been drawn to wildlife in Nagarahole National Park, not far from his home in Mysore, and he mixed easily with wildlife biologists. In due course he gained admission to the University of Florida master's program, where he planned his research with Mel Sunquist, who knew McDougal from his own studies of tigers in Chitwan National Park a few years earlier.

Impressed with McDougal's camerawork, Karanth settled on camera traps as one of his research tools in 1986. However, the few expensive single-lens reflex cameras used by McDougal did not meet Karanth's need for sampling tiger populations with a large number of cheap cameras that were easy to deploy. Early in the 1990s he finally found his tool in TrailMaster camera traps.

MCDOUGAL'S LESSONS LEARNED

Figure 3.7. Chuck McDougal, an anthropologist by training, began camera trapping tigers in 1974.

Chuck McDougal's earliest camera trap, a Nikon F2, was ideally suited for use with external or remote triggers by virtue of its accessory electronic shutter. He built a pressure plate from two wooden planks and used compression springs to hold the electrical contact of the switch apart. He buried the pressure pad and its cable on the trail and whisked the sandy soil smooth. When an animal stepped on the wooden pad, the battery-powered circuit was closed, instantly triggering camera and flash.

McDougal steadfastly improved the system to capture stunning images taken at eye level. By placing a slave-activated flash at a 45-degree angle to the trail, he eliminated the flat appearance rendered by head-on lighting of the camera's electronic flash.

I had the pleasure of joining McDougal in 1981 as he set his camera on a trail near Tiger Tops, but I was baffled when he dressed the camera in two beautifully crafted black quilted jackets with an opening for the lens.

"Why make the camera look like a black scarecrow?" I asked. They were mufflers, he explained, made by the local tailor and stuffed with seedy fiber from the local silk cotton tree. Like locally made pillows, the muffler was highly attractive to rats with an appetite for oily seeds. But they solved a vexing problem. In pre-muffler days tigers were often photographed in unflattering moments, frozen in motion veering away from the camera, or with the unmistakable "possum in the headlights" look. The big cats reacted within microseconds to the sound of the camera's shutter, testimony to the tiger's lightning reflexes. As for the pillow-headed scarecrow on the side of the trail, the big cats didn't seem to notice or care.

In the mid-1990s McDougal started using TrailMaster monitors. He set the first one for three days on the ridge behind Tiger Tops.

> When I went to collect it, I found that the tripod had been knocked over and the camera was missing. A young but large male tiger had attacked and carried the camera in his jaws about 80 meters down the path before dropping it. The slide had been knocked off and the camera had canine dents in it. But the photo of the tiger, giving the camera a dirty look before attacking it, came out fine. I sent it to Bill Goodson who put it on his website and gave me a new camera.[23]

Over the years McDougal photo-captured an impressive inventory of other wildlife, including yellow-throated martens, several species of civets, and striped hyenas. The rarest find perhaps was a **ratel**, or honey badger, now a rare species in the Indian subcontinent.

Since tigers are not marked identically on both sides of their bodies, Karanth began using the TrailMaster system to photograph both sides of the animal simultaneously with two cameras (Figure 3.8*B*). However, even after a number of tigers were identified and counted, a key question remained: How many tigers were not caught in the camera traps and thus omitted from the count?

Figure 3.8A. Ullas Karanth's camera trap studies of tigers in India ushered in new developments and statistical methods for estimating population size of individually identifiable mammals.
Courtesy of Ullas Karanth

As data accumulated, Karanth realized he needed a statistical ecologist as a collaborator, so he joined forces with American Jim Nichols of the US Fish and Wildlife Service. Together they developed state-of-the-art statistical models and field practices based on well-established capture–recapture models.[22] Camera traps made it possible to accurately count not just tigers but many other elusive species in which individual recognition is possible. Their approach now enables camera trappers to estimate numbers of such creatures as leopards, snow leopards, jaguars, and even elephants.

CAMERA TRAPS FOR ANYONE AND EVERYONE

Electronic camera technology was rapidly advancing by the mid-1980s, when Japanese-made point-and-shoot cameras reached new levels of sophistication. With automatic exposure and film advance, autofocus lenses, and built-in electronic flash, new compact cameras were just a step away from a completely remote camera trap. All that was needed was a sensor to detect wildlife and circuitry to power the camera and trigger the shutter. Trip wires, baited switches, and pressure pads didn't answer the need for a simple and effective sensor.

When a Kansan named Bill Goodson killed a white-tailed buck exactly where he had seen it the day before, he had an embarrassingly obvious revelation: Deer are creatures of habit. He also realized he could make an electronic aid that could tell hunters where the deer are. With a background in everything from digital communications to radar and missile guidance systems, Goodson knew

Figure 3.8*B*. Karanth used two cameras to photograph both sides of the body because every tiger is distinctively marked and both sides of the body are different.
Courtesy of Ullas Karanth

Figure 3.9 By the 1980s Bill Goodson's TrailMaster products were popular among many users, including wildlife biologists. *Courtesy of Bill Goodson*)

electronics (Figure 3.9). He also knew that recent advances in thermal detection were well suited for remote monitoring of wildlife.[24]

In 1981 he started experimenting with passive and **active infrared (AIR)** detectors, made prototypes, and talked them up and sold them at hunting and shooting shows. Five years later he unveiled the TrailMaster infrared trail monitoring system, the first consumer product that coupled infrared sensing capability with electronic counting circuits and cameras. It was mainly a hunter's aid, but Goodson also wanted to appeal to more technical **niche** markets. A data-retrieval device and statistical software, for example, addressed the needs of biologists and park managers.[25]

Goodson's passive and AIR monitors range in versatility from simple electronic game counters to programmable devices with multiple functions. **Passive infrared (PIR)** trail monitors have wide-angle detection ranges as far as 65 feet, while AIR monitors use a transmitter and receiver to target a particular species based on the average duration of its infrared event. Both can be programmed to count events and specify delays between events, and they can signal one or more cameras to take photos or video clips.

In 1988 Goodson integrated the AIR monitors with a point-and-shoot camera and a video camera. The Olympus 35mm AF-1 Twin Quartz Date was weatherproof, had two lenses, and had a surprisingly quick **reaction time**. The Sony Handycam Video8 allowed first-class video footage. The system was also compatible with digital single lens reflex (DSLR) cameras.

Goodson's products answered both the general needs of hunters and photographers and the technical needs of wildlife biologists wishing to monitor wildlife and park visitor activity.

THE RISE OF COMMERCIAL TRAIL CAMERAS

With millions of deer killed annually by North American hunters, other engineers and entrepreneurs heeded the call of a potentially lucrative market for dedicated trail cameras. In 1989 new start-ups appeared in two states with strong hunting traditions. In western Georgia, CamTrakker Inc. started producing a commercial 35mm point-and-shoot camera integrated with a controller circuit and a PIR sensor. In Green Bay, Wisconsin, Cuddeback started selling a trail camera using circuit-board cameras fitted to a plastic case. Digital cameras quickly supplanted 35mm film cameras as dedicated trail cameras came into being. The race was on.

In the 1990s independent and established outdoor equipment companies began producing their own digital trail, scouting, and game cameras. By the turn of the century over two dozen companies, mostly in the United States, competed for market share. Most of them turned to the Chinese industrial sector, where production costs were significantly less than in the United States.

The nonhunting public quickly saw the utility of the new products. They were accessible and affordable, and they offered naturalists an intriguing new way to observe wildlife. Field biologists adopted them for wildlife research, and outdoors enthusiasts bought them for observing wildlife. But many buyers found that the early digital trail cameras were unremarkable photographic instruments and voiced their opinions on hunting and outdoor forums and in retailers' online product reviews. Independent trail-camera reviewers subjected the new products to technical tests and posted the results as a service to consumers.

It became apparent that most customers wanted trail cameras that could detect animals near and far and photograph them before they moved out of view. They had limited tolerance for blank exposures; it didn't matter if the blanks resulted from a tardy shutter or a shutter tricked by thermal noise from moving vegetation. On the other hand hunters seemed more tolerant of modest image quality than other users. Photographers seeking high-quality images simply stuck with DSLRs and quick triggers like TrailMaster's active infrared monitors.

Consumer desires for improvements spurred competition, and manufacturers reduced or eliminated the deficiencies in successive models. Rapid **trigger speed**, long detection distance, and corresponding reach in infrared illumination became key selling points. Image **resolution** improved with better **image sensors** and by **interpolation**. Added capabilities included improved power efficiency, solar chargeable batteries, video, **time-lapse**, and Wi-Fi. Temperature and moon phase were added to time stamps. Consumers who were pleased with some particular feature were often soured to find it missing in the next model. Bells and whistles came and went, but the products gradually improved.

Despite continuing advancements, some enthusiasts believed that commercial trail cameras were overpriced, and others wanted more from them than the industry was willing to give. This fringe group was a niche market that wasn't being served, and when industry does not respond to a felt need, innovators answer the call.[26]

REBELLIOUS HACKERS

Desire for a better trail camera at lower cost energized some gifted do-it-yourselfers to make their own. These rebels opened a few models of 35mm point-and-shoot cameras, located power, shutter, and ground terminals, and wired them to homemade controller circuits and PIR circuits from home security lights. DIY instructions soon appeared on *Jesse's Hunting Page* with bills of goods, tools, price estimates, and warnings to would-be tinkerers that opening the camera voids the warranty. The idea appealed to others of like mind, and the new community grew in both size and scope. Someone associated with the movement saw a similarity between home-brewing beer and hacking trail cameras at home. Trail cameras hacked and built at home became known as *home-brewed* cameras or simply *homebrews*.

The **homebrew** movement created a need for electronic components, and a handful of electronic engineers, programmers, and businessmen, usually moonlighters, started to sell dedicated homebrew controllers with built-in PIR sensors and related supplies. Online chat groups and YouTube videos advanced the cause by sharing advice, personal experiences, hacking tutorials, and photos.

Homebrewers soon discovered that point-and-shoot cameras differed greatly in trigger speed. Slow triggering of photos following sensor detection meant missed or partial animal images. Online forums and homebrew suppliers tested new models in a search for "fast cameras." In the

Sony s600, an all-time homebrewer's favorite, engineers disabled the lens cover and locked the extendable lens in a state of picture readiness. This modification, known as a fast hack, decreased shutter reaction time by a second.

Homebrewers have had to accommodate ongoing changes in point-and-shoot design. When larger telescopic lenses became popular, hackers searched for suitable tubes so that lenses could extend beyond the front of the camera housing. Then customized tubes, **snorkels** in homebrew parlance, became available on homebrew supply sites. The weak flash of most point-and-shoot cameras demanded the use of **remote flashes** as part of the build.

The driving force for homebrew development has been photographic versatility, high-quality photos, and ease of conversion for remote use. Wildlife photographers and researchers often require customized equipment for specialized applications. The demand for smart phones now detracts from the point and shoot camera market, and camera trappers increasingly seek DSLRs for their work.[27]

To use a biological metaphor, camera trap design is in the midst of an adaptive radiation. The commercial trail camera industry is the big branch on this evolutionary tree that supplies the general needs of a diverse group of outdoor recreationists. The homebrewing community is the small branch that addresses more specialized needs and interests in camera building and photography. They are not mutually exclusive groups. Camera trappers jump from branch to branch as it suits them.

MOTIVES AND ETHICAL UNDERPINNINGS

In an exhaustive review of trail cameras as research tools, biologists Tom Kucera and Reginald Barrett wrote: "Combining human curiosity and ingenuity, . . . remote camera techniques have allowed previously unimaginable access into the lives of many wildlife species."[28] The soaring growth of camera trap publications validates their claim. Besides scientific benefits, trail camera images of rarely seen and rediscovered species have drawn public attention to the thrills of zoological discovery.

Pursuing curiosity with remote cameras has always evinced a certain degree of reflection on nature. Pioneer camera trappers, usually reformed hunters, felt compelled to proclaim the ethical dividends of shooting animals with cameras rather than guns. Their images of wildlife were often evocative statements about untrammeled nature. Their contemporaries who hunted may have shared their sense of kinship with nature, but they certainly didn't concur with their condemnation of hunting. Cameras did not put food on the table; they were luxuries and tools of the elite. Hunters valued wildlife mainly for its material utility, while camera trappers reveled in its esthetic and spiritual rewards.

Second-generation camera trappers, like the pioneers of the set camera, were naturalists who also built their own automatic cameras. But it wasn't as easily done as before—cameras had become more complicated, and field biologists wanted more from them. Zoology had also grown more quantitative, and biologists needed cameras that shuttered photos repeatedly. These constraints put camera traps into the hands of biologists who had compelling needs for them and the faculty to build them. Their motives were scientific, and though creating artistic photos was not a goal, they were not without esthetic sensibility. Many of them would confess that a species' subjective appeal influenced their choice of study topic.

Mass production of trail cameras introduced affordable camera trapping to a large segment of society with assorted attitudes toward and interests in nature. No longer was the remote camera a tool that naturalists and field biologists had to fabricate for their exclusive use. Hunters, wildlife biologists, amateur naturalists, and sundry outdoor recreationists are the third generation of camera trappers who share in the tangible thrills of seeing the unseen. The hunters finally had a tool to survey their quarry and learned that shooting wildlife with a camera could become a gratifying end in itself. Nonhunters camera-trapped to satisfy their dormant curiosity in nature. Whatever one's reasons for having them, trail cameras have an inherent capacity to arouse people's curiosity in nature, raise questions, and occasionally lead to biological discoveries.

CHAPTER 4

Camera Trapping Discoveries

Camera traps have ushered in the discovery of species new to science and unveiled the existence of species thought to be long extinct. They have also documented the presence of species beyond the boundaries of their accepted geographic distribution. Camera trap users continue to surprise us with photos and clips of mammals and birds displaying seldom seen skills and habits. Here I tell a few stories of discovery made by biologists and naturalists while pursuing research or recreation in wild places.

Figure 4.1. Francesco Rovero tends camera traps with assistants in the Udzungwa Mountains of northern Tanzania. *Courtesy of Francesco Rovero*

In 2005 Francesco Rovero got the thrill of a lifetime. The Italian zoologist was reviewing camera trap photos in the Udzungwa Mountains of northern Tanzania when the image of a sengi, or elephant shrew, popped up on the **LCD** (Figure 4.1). Sengis are very different creatures from the secretive forest antelopes called **duikers** that Rovero was studying.[1] Sengis are bizarre members of an ancient lineage and might have been the real-life inspiration for the fictional snouters created by the late German professor Gerolf Steiner.[2] Their nearest relatives include a strange mix of improbable wonders—elephants, sea cows, hyraxes, and aardvarks.

This particular sengi was garishly marked and not to be found in any field guide of East African mammals (Figure 4.2). It had a grizzled face with a cream-colored chin, a yellow breast, orange sides, and a maroon back. Its hindquarters were jet-black.[3]

Figure 4.2. Rovero's garishly colored new species was christened the gray-faced sengi *(Rhynchocyon udzungwensis)*. *Courtesy of Francesco Rovero*

Rovero knew he had found something unusual. He contacted Galen Rathbun, an American who had been studying sengis since his Peace Corps days in Kenya in the 1970s. Rathbun verified Rovero's suspicion, but they needed physical evidence to prove the discovery. After all this sengi might only be a colorful race of an existing species. A year later they mounted an expedition to gather specimens and details of natural history.

Back in civilization they compared the specimens with those in museums and confirmed their expectations: the sengi was new to science. They named it *Rhynchocyon udzungwensis*. At 700 grams, it was the size of a small cat and among the largest sengis known to science.

As I write this account, the gray-faced sengi has the distinction of being the only new species of mammal discovered with a camera trap. But it will not hold that status forever. At least two other species have been discovered with camera traps and are in the queue awaiting scientific description.

THE SAOLA

Nowadays biologists encounter new species of animals for sale in rural markets or find their remains in the possession of hunters. Then they use camera traps to try to define the animals' geographic range and habitat. The discovery of the saola is an example.[4]

The story began in 1992 in the Annamite Mountains along the Laos–Vietnam border. There, in a village house, a joint team of Vietnamese biologists led by conservationists John MacKinnon and Vu Van Dung found a hunting trophy—a baffling skullcap with horns (Figure 4.3*A*). As a seasoned expert on Asian wildlife, MacKinnon knew the remains were unique and unknown to science, but the saola was no secret to local hunters. How could an ungulate of this size have escaped the notice of western naturalists during the French colonial period? In fact the hill tribes that lived in those forbidding jungles vigorously rebuffed French and ethnic Vietnamese intruders alike.[5]

The team reported the finding to forestry authorities, and a few months later MacKinnon led a team from the Vietnam Ministry of Forestry and

Figure 4.3A. MacKinnon with the first two skull caps of saola found during the first expedition. *Courtesy of John MacKinnon*

World Wildlife Fund for Nature (WWF) to gather more skulls and records (Figure 4.3*B*). The expedition soon learned the reason for the saola's late discovery: it was a rare species that lived in isolated and forbidding forests. They published their description of the new species in 1993.[6]

Figure 4.3*B*. *Left to right:* Do Tuoc, John MacKinnon, and Vu Van Dung pose with a taxidermy mount of a saola *(Pseudoryx nghetinhensis)*. It was reconstructed from the first intact hide recovered in the Annamite Mountains, Lao–Vietnam border region.
Courtesy of John MacKinnon

Hunters captured several animals in the ensuing years, but none of the docile captives lived longer than five months. One of these was the subject of detailed observations published by William Robichaud, an American wildlife biologist working for the Wildlife Conservation Society.[7] He trained villagers in Ban Vangban village to use camera traps, and they photographed the first wild saola in 1999. Since then saola have been camera-trapped five times—twice from Bolikhamxay province in Laos in the late 1990s, twice from Pu Mat National Park in Vietnam in the late 1990s, and once from central Vietnam in 2013. According to Andrew Tilker, Asian species officer at Re:wild, "The lack of records since then is not for lack of trying: There have been hundreds of thousands of camera trap nights across the Annamites in the past [approximately] fifteen years—indicating that the saola is likely very rare (probably a handful of individuals, if we are lucky)."[8]

RABBITS WITH TIGER STRIPES

Photos of living animals stir the public's imagination, and camera trap photos of new or rediscovered species have popularized zoological fieldwork. Photos of dried specimens in a museum cabinet simply don't capture the essence of the living animal.

In the mid-1990s British biologist Robert Timmins, an employee of the New York-based Wildlife Conservation Society, found a hunted rabbit for sale in a rural market in central Laos. The country bazaar is a busy crack-of-dawn phenomenon where farmers, hunters, and gatherers assemble to sell their wares from far and wide. Timmins knew the outrageously marked rabbits were not of a domestic breed. Their stripes made them look like wannabe tigers (Figure 4.4).

This was a thrilling find because the only species of striped rabbit known to science, the Sumatran striped rabbit (*Nesolagus netscheri*), was found only from the mountains of Sumatra 1,500 kilometers away.[9] Timmins encountered more specimens in markets elsewhere in the Annamite Mountains, and he and his colleagues proved the animals to be a new species, which was named *Nesolagus timminsi* (Annamite striped rabbit) in honor of its inquisitive discoverer.[10] The two species likely evolved from a common ancestor when contiguous Southeast Asia, or Sundaland, became the Indonesian archipelago.

Figure 4.4. This Annamite striped rabbit *(Nesolagus timminsi)*, photographed using a handheld camera, is a good example of disruptive coloration. (Hue SNR)
Courtesy of Andrew Tilker

Three years passed before staff of the British charity Fauna & Flora International took the first photos of living Annamite striped rabbits in Vietnam with camera traps.[11] Despite sporadic camera trapping in Laos, the new species wasn't camera-trapped there until 2007, when a Laotian team trained by the Wildlife Conservation Society found it in Nakai-Nam Theun National Protected Area. Records from protected areas in Vietnam and Laos showed that the striped rabbit, like the saola, was an endemic species of the Annamite Mountains.[12]

Annamite striped rabbit discoveries have continued to this day. In 2020 biologists got camera-trap photos of the Annamite striped rabbit in Vietnam—more than 400 kilometers from the nearest known population. The researchers were shocked, to say the least. In terms of biogeography, finding Annamite striped rabbit in southern Vietnam was a bit like finding a polar bear in the American Midwest.[13]

The Sumatran striped rabbit was also rediscovered in 1998, making it the year of the striped rabbit. Jeremy Holden of Fauna & Flora International accidentally camera-trapped an animal in Kerinci Seblat National Park. The Sumatran rabbit became known to western science in 1880 when Dutch naturalists collected specimens in the mountains of western Sumatra, but it hadn't been seen or collected since 1929.[14] These were great findings for conservationists, and the camera trap photos bolstered the case for protecting the remaining habitats of these rare and unusual species.

Figure 4.5. A Papuan woolly rat *(Mallomys sp.)* poses with cameraman Gordon Buchanan on Mount Bosavi, Papua New Guinea.
Courtesy of Jonny Keeling, BBC Natural History Unit

A BEHEMOTH AND A METHUSELAH AMONG RATS

The Bosavi woolly rat (*Mallomys* sp.) is a rodent that has new species written all over it. The animal is a behemoth among rats, weighing about three pounds and measuring close to three feet from head to tail (Figure 4.5). It was discovered in 2009 in the 3,000-foot-deep forested crater of Mount Bosavi, Papua New Guinea, where a BBC team was filming *The Lost Land of the Volcano*. Smithsonian mammalogist Kris Helgen collected specimens and recognized it as an undescribed species, and wildlife photographer Gordon Buchanan photographed it with a TrailMaster AIR sensor and a digital camera. The film, and the discovery of the giant rat, won the team instant kudos, even though the remarkable rat remains to be scientifically described.[15]

Figure 4.6. A wild kha-nyou *(Laonastes aenigmamus)* photographed in daylight in Phou Hin Poun National Biodiversity Conservation Area, Laos. The species was discovered for sale in a country market and proved to be the only member of an extinct family of rodents. *Courtesy of Tilo Nadler*

The Laotian rock rat, or kha-nyou, on the other hand, doesn't look quite like a rat. If anything, it looks like a degenerate squirrel, with small eyes, a long head, and a furry tail like a chipmunk's (Figure 4.6).

Once again it was Robert Timmins, the inveterate habitué of rural markets, who discovered and brought the kha-nyou to the attention of western science. He found the strange rodents for sale next to the fresh vegetables on an early morning visit to a Laotian bazaar. The rodents have probably long been a staple in the villagers' diet.

The sight of dead rats would have creeped out most western tourists, but Timmins is an Asia Hand who knows the fauna, and these rodents were unlike anything he had seen before. He learned they were caught nearby in rugged limestone mountains, and he bought the specimens—not to eat, but to place in the hands of specialists.

The wheels of science often turn slowly, and nearly a decade passed before the strange rodent was scientifically described as *Laonastes aenigmamus*, which translates to "enigmatic rock rat."[16]

Modern DNA analyses often uncover new species among old specimens in museum cases, but the media rarely deem a new species of rat newsworthy. The kha-nyou, however, was weird and wonderful because it didn't fit into any family of living rodents. The scientists gave it its own family: the Laonastidae.[17]

A new family of mammals is certainly newsworthy, and other scientists were soon champing at the bit to behold the enigmatic rat. One of them was an American paleontologist named Mary Dawson, who rallied colleagues to join her in examining the specimens. They recognized the kha-nyou's similarity to an extinct family of rodents and concluded that it is the sole surviving species of an ancient family known as the Diatomyidae.[18]

Their findings, reported in the paper "*Laonastes* and the 'Lazarus Effect' in Recent Mammals," published in *Science* in May 2006, stirred global interest. Though the rock rat's nearest relatives

Figure 4.7. Artist Inge van Noortwijk's sketch of the "mysterious red furry creature" from Indonesian Borneo helped scientists conclude that the camera trapped animal was most probably a tree squirrel. *Courtesy of Inge van Noortwijk, Naturalis Biodiversity Center Leiden, Netherlands*

had gone extinct eleven million years ago, this species—ensconced in an isolated limestone karst bordering Laos and Vietnam—had mysteriously escaped the notice of western science. Indeed it was a living fossil that, like Lazarus, had returned seemingly from the dead.[19]

The published photos of the remarkable kha-nyou's corpses didn't do justice to the magic of its scientific resurrection. At least a few people wanted to see a perky kha-nyou with sparkling eyes, and one of them was David Redfield, a retired professor and science educator from Florida. Rare mammals captured the professor's imagination and photographing them had become his cause. The rock rat was worthy quarry. Redfield invited his friend Uthai Treesucon, a Thai wildlife biologist, to join him on an expedition to Laos, and two months later they embarked on their quest to photograph the living fossil.[20]

Diplomacy won their acceptance by local hunters, who captured a live rock rat after several failed attempts. The video footage shows a darkly handsome rodent meekly crawling on Treesucon's arms and shoulders. It was anything but perky.

The kha-nyou might have been discovered with a camera trap. Rodents aren't camera shy, and most can be baited if you know where to look for them. Commercial trail cameras, however, are not often designed to photograph small animals, and identifying species from such images often tests the imagination, especially in locations where the fauna are not well known. Show a specialist a trail-camera picture of an unusual rodent or shrew from such a place, and they will study it with foreboding frustration. When they ask "Did you collect a specimen?" you will feel like the kid who wasn't listening to the teacher. Under certain conditions, trail cameras have a way of creating images that become unsolvable mysteries.

THE BIZARRE AND UNIDENTIFIABLE

In 2003 biologists from WWF Switzerland captured two camera trap images of "a mysterious red furry creature" in Kayan Mentarang National Park in Kalimantan, in Indonesian Borneo. The images were simply puzzling, even to local villagers. WWF consulted mammal specialists, who suggested they try to catch the animal. They tried in vain.[21]

The excitement of the biologists was understandable. Borneo is a hotspot of biodiversity, the kind of place where scientists find new species. WWF decided the new species would serve a worthy cause to protect this forest and others like it and released the images in 2005 with the publication of a book on tropical forests.

The wonderment of the media led to tantalizing speculation: It was "slightly larger than a domestic cat;" it resembled a marten or civet; and it "looks like it is half cat and half fox." Someone speculated that it "may belong to an entirely new group." WWF experts said: "This could be the first time in more than a century that a new carnivore has been discovered on the island."

Scientists and researchers offered their opinions. Could it be a range extension of the rare Hose's civet known only from northern (Malaysian) Borneo? Or could it be an unknown population of the strange, catlike fossa from Madagascar?[22]

Three biologists decided it was time for a more analytical approach. Comparing the features with seventeen species known from the region, they concluded that the mystery mammal was probably a flying squirrel, either Thomas's flying squirrel (*Aeromys thomasi*) or the red giant flying squirrel (*Petaurista petaurista*).[23] It took an artist's rendition of the photo to uncloak the mystery mammal as a squirrel (Figure 4.7). That ended the storm of imagination surrounding the discovery of the "mysterious red furry creature" of Borneo.

Figure 4.8. Camera trap photos of the silver-backed chevrotain (*Tragulus versicolor*) confirmed its continued existence in Vietnam and hastened conservation action. (Nui Chua National Park, Vietnam)
Courtesy of Re:wild, Southern Institute of Ecology, Leibniz Institute for Zoo and Wildlife Research

REDISCOVERING LOST AND LEGENDARY SPECIES

The most common kind of camera trap discovery is actually rediscovery of animals that have been forgotten or lost from recent human memory. The media are prone to describe these animals as *lost* or *legendary* species. Some are truly rare species, like the hairy-nosed otter (*Lutra sumatrana*). This white-throated otter with an unusually hirsute nose is known from scattered localities in Southeast Asia and was thought to be extinct in Sumatra, where it was originally discovered. The discovery of road-killed animals in Borneo and southern Sumatra in 1997 and 2005 kindled hope that it still existed, and WWF confirmed its existence when its staff camera-trapped one in 2008.[24]

There are other examples. In 2008 Debbie Martyr of Flora & Fauna International found a rare Sumatran **muntjac** (*Muntiacus montanus*) in a poacher's snare. It was thought to be extinct, but this was proof that it still existed. Martyr confirmed its presence using camera traps.[25] Likewise, it was an exciting find when field workers from Iran's Department of Environment and the Wildlife Conservation Society got camera trap photos of a family of five cheetahs in 2005. Before then it was thought that cheetahs were extinct in Asia.[26]

The Sumatran ground cuckoo, critically endangered and known only from eight specimens and a few sightings, was camera-trapped in 2006 to the delight of the ornithological community.[27]

In 2017 a team of wildlife conservationists confirmed the existence of the silver-backed **chevrotain** (*Tragulus versicolor*) in Vietnam (Figure 4.8). Though it was scientifically named and described in 1910, the diminutive ungulate remained a mystery to western science until a

Figure 4.9. Photographic evidence from Vancouver Island that a hungry turkey vulture (*Cathartes aura*) will eat the unthinkable—the feces of a mountain lion.
Courtesy of Gary Schroyen

Vietnamese–Russian survey team laid hands on a hunter-killed specimen in 1990. Nearly three decades passed before another team of wildlife biologists verified its continued existence. Their camera trap sets in several areas recorded nearly three hundred visitations and more than 1,800 photos of the species.[28]

Camera traps prove that species of doubtful existence are still with us and justify actions for their conservation.

THRILLS OF PERSONAL DISCOVERY

Biologists have learned a great deal from field studies using trail cameras, and the proof is in the ever-growing body of scientific literature.[29] Recreational camera trappers have also gained insights into the ecology and behavior of their local species. Their findings may not be new to science, but that doesn't detract from the personal satisfaction of discovery they garner. Firsthand experiences always trump virtual ones, and most discoveries raise more questions than they answer, but amateur naturalists have gained important insights and written about them in forums, websites, and blogs.

The turkey vulture's miraculous power of smell was a remarkable scientific discovery.[30] Evidence of this faculty continues to amaze camera trappers, as it did Gary Schroyen when he viewed several images of one sampling a recent mountain lion scat in Vancouver Island (Figure 4.9). The site was hidden by a closed canopy of conifers, so it is likely the bird probably scented the scat while riding thermals overhead. Schroyen's camera captured convincing evidence that turkey vultures at times ingest carnivore feces.

It's a pleasant surprise to capture a photo of a swooping bat. Bats are most often recorded at roosts or water sets, but on summer nights near Duluth, Minnesota, Sean Hall has camera-trapped bats swooping on deer. More recently, Daryl Dean captured a bat swooping near a moose (Figure 4.10A-C). Coincidence, or is something going on? Deer often attract clouds of mosquitoes, and the obvious conclusion is that bats feed on the mossies that are feeding on the deer.[31] Observations like these raise questions. Can bats discriminate between nutritious, blood-gorged mosquitoes and skinny, hungry mosquitoes? And is this how obligate blood feeding evolved in vampire bats?

"The bat from hell" is the most popular film clip I have posted to YouTube.[32] This small bat, probably a species of ***Myotis***, strafed a western screech owl standing at the edge of a seasonal creek. The bat appeared to have actually grazed the feathers on the owl's head, and the owl immediately took flight. Was this an accidental collision by a sonar-impaired bat? Was it gleaning a mosquito from the owl's head? I shared the video with a

Figure 4.10A. Large mammals seem to be defenseless to swarms of mosquitoes.
Courtesy of Chuck Gackstetter

Figure 4.10. *B*, Camera traps have garnered circumstantial evidence that bats glean mosquitoes from white-tailed deer in Minnesota. *Courtesy of Sean Hall.* *C*, Bats also seem to associate with moose and their associated mosquitoes in Saskatchewan, Canada. *Courtesy of Daryl Dean*

bat specialist, who thought I had captured a rare event of mobbing behavior, a predator harassment phenomenon well known to suburbanites who have been dive-bombed by nesting birds. Tropical bats are known to mob predators, but little has been written about mobbing by insectivorous bats.[33]

Mobbing behavior also explained video footage of several *Myotis* hovering over a great horned owl about to bathe in a creek. The bats were clearly a distraction, and the owl got on with its bath only after the mobsters left.[34] These findings were made after months of camera trapping the same pools in a seasonal creek, proving that patience occasionally pays off.

Ken Hickman's *Nature of a Man* is a nature blog about personal discovery written with contagious enthusiasm. Most posts are about his adventures in camera trapping. Few camera trappers regard small rodents as engaging subjects, but Hickman monitored numerous stick nests of woodrats, which he describes as being "built like castles of yore in-and-around solid natural features." Nests and dens are centers of activity, and by systematically working nests in several locations, he mined the area for all it was worth.[35]

Figure 4.11. These highly cropped images were good enough to identify the African striped weasel *(Poecilogale albinucha)* in the Tygerberg Nature Reserve—where it wasn't "supposed to be." *Courtesy of Henry de Lange*

Hickman was able to identify the members of the woodrat community. He documented food plants, encounters with neighbors, transport of young, and a long list of visiting mammals and birds, including nest **commensals** and three species of salamander. His interpretations of his photos are based on the technical literature and are a fine example of how personal discovery leads the curious naturalist to the biological literature and even to delve into research.

Recreational camera trappers may also make discoveries unknown to science. That is what happened when Henry de Lange, a South African IT specialist, camera-trapped a streamlined black-and-white mammal at Tygerberg Nature Reserve in the northern suburbs of Cape Town.[36] He identified it in a field guide as a striped weasel (*Poecilogale albinucha*) but learned that

Figure 4.12. Officials believed de Lange had misidentified the striped weasel for the honey badger *(Mellivora capensis)*, *left*, and the striped polecat *(Ictonyx striatus)*, *right*, which were both known to exist in the reserve. He used these images to show the differences. (Sony s600)
Courtesy of Henry de Lange

they were not known to be in this part of South Africa (Figure 4.11). His photo seemed to document an extension of the weasel's geographic range, so he sent it to the reserve's administrators with a letter explaining the discovery.

The officials had reason to doubt the identification because other black-and-white mammals—the honey badger (*Mellivora capensis*) and the striped polecat (*Ictonyx striatus*)—were known to exist in the reserve (Figure 4.12). But de Lange knew that his photos contained the proof because he had also camera-trapped these other species. He defended his identification point by point, listing the differences between the three species. To add the perspective of body size he overlaid the weasel's image with those of rodents and a robin-chat taken at the same set. The weasel was clearly a sprite compared to the ratel and polecat. This ended the friendly debate. It was an uphill battle, but the skeptics finally accepted the validity of the discovery, which proves that even recreational camera trappers can make scientific discoveries.

CHAPTER 5

What Is a Camera Trap?

Camera traps are automatic battery-powered cameras triggered with a sensor and are intended for remote out-of-door use. They include a wide variety of ready-to-use commercial products that have **fixed-focus lenses** and relatively low-resolution image sensors. Homemade (often called *homebrewed*) camera traps are usually constructed from DSLR and point-and-shoot cameras with additional electronic components. They produce higher quality images than **commercial trail cameras**. Both commercial and homemade cameras take still photos and video clips in color by day, as well as time-lapse. Most commercial trail cameras take infrared night video, but few homebrewed cameras can be hacked to take infrared videos at night. This chapter describes critical camera operations and settings, as well as various options for building camera traps.

Camera traps are also known as trail cameras, scouting cameras, game cameras, critter cams, and even nature cameras.[1] Regardless of the name, they all operate in the same way. The photographer selects a suitable place, programs the settings, sets the camera, and leaves the scene. Electronic circuits do the rest. Camera traps are intended as remote devices to be left in isolated locations, often exposed to the elements for days or months at a time.

Camera trapping had relatively few practitioners for nearly a century, but with the advent of commercial trail cameras in the 1980s, it became a popular pastime. Buyers now face the dilemma of deciding which of the numerous models suits their needs and budget. Reading product reviews on the internet[2] can help buyers make informed decisions, and other online resources can help those who want to build a camera trap from parts (see "Camera Trapping Forums" and "Camera Trap Building Supplies and Services" later in the chapter).

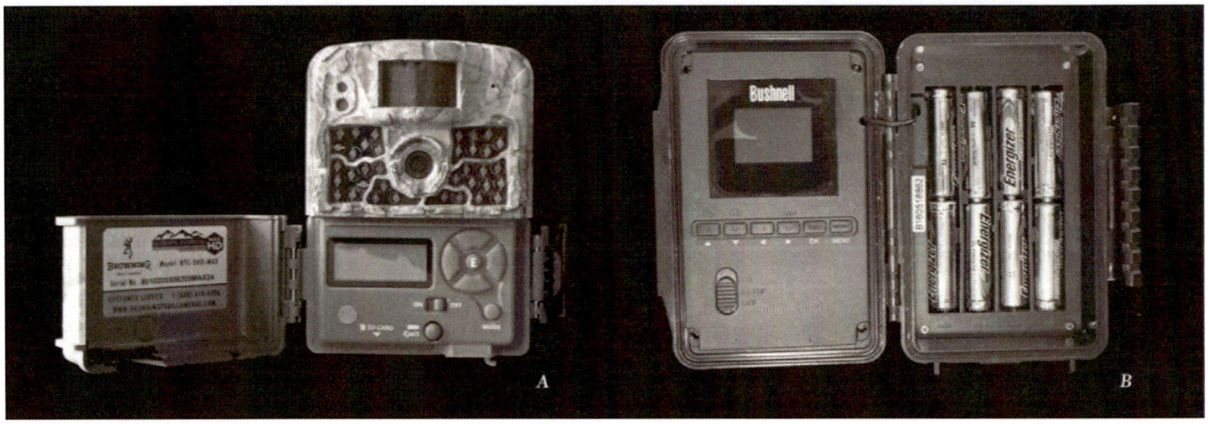

Figure 5.1. Though commercial trail cameras share basic features, they differ in design and operational details. ***A***, Early and more economic models, like Browning's Strike Force Max HD, feature small LCDs used solely for set-up. ***B***, The Browning Patriot has a color LCD dedicated to both set up and view finding; it features separate lenses for day and night shooting. Research the options and read reviews before you buy a new camera. There are many makes and models available.

This chapter explains the general features and operation of camera traps and describes the differences between commercial and homemade ones.

A QUICK COMPARISON OF COMMERCIAL AND HOMEBREWED CAMERA TRAPS

Commercial Trail Camera Traps

Modern camera traps are either commercial products or handmade devices assembled from various components. Commercial trail cameras are self-contained units designed for outdoor recreation (Figure 5.1). They are built around a small digital camera module consisting of a lens on a **printed circuit board**. Their fixed-focus lenses give sharp images between 10 feet and infinity, and their lenses have a relatively narrow (between 35 and 55 degrees) **field of view (FOV)**. A few models feature wide-angle lenses (between 110 and 120 degrees).

The flash is usually an infrared **LED** array, and a PIR sensor detects animal movement. Most models have an LCD screen for the set-up menu and for viewing in shoot and picture modes. Still images and video clips are standard capabilities, and time-lapse is becoming increasingly popular. In fact a few models take only video or time-lapse pictures. The ability to automatically transmit photos to a cellular phone or computer is another recent advancement. Trail cameras have made great strides in improving photographic quality, and the market offers a wide range of choices. This has converted some homebrew advocates to the use of commercial trail cameras, but many camera trappers are devoted to the photographic versatility of homebrewed cameras, DSLRs in particular.

Homebrewed Camera Traps

Homebrewed camera traps are built from a variety of cameras and components, such as controllers, cases, and external flash or infrared arrays. These cameras have more options than commercial trail cameras, and they can be disconnected and used out of the case for snap shots or video.

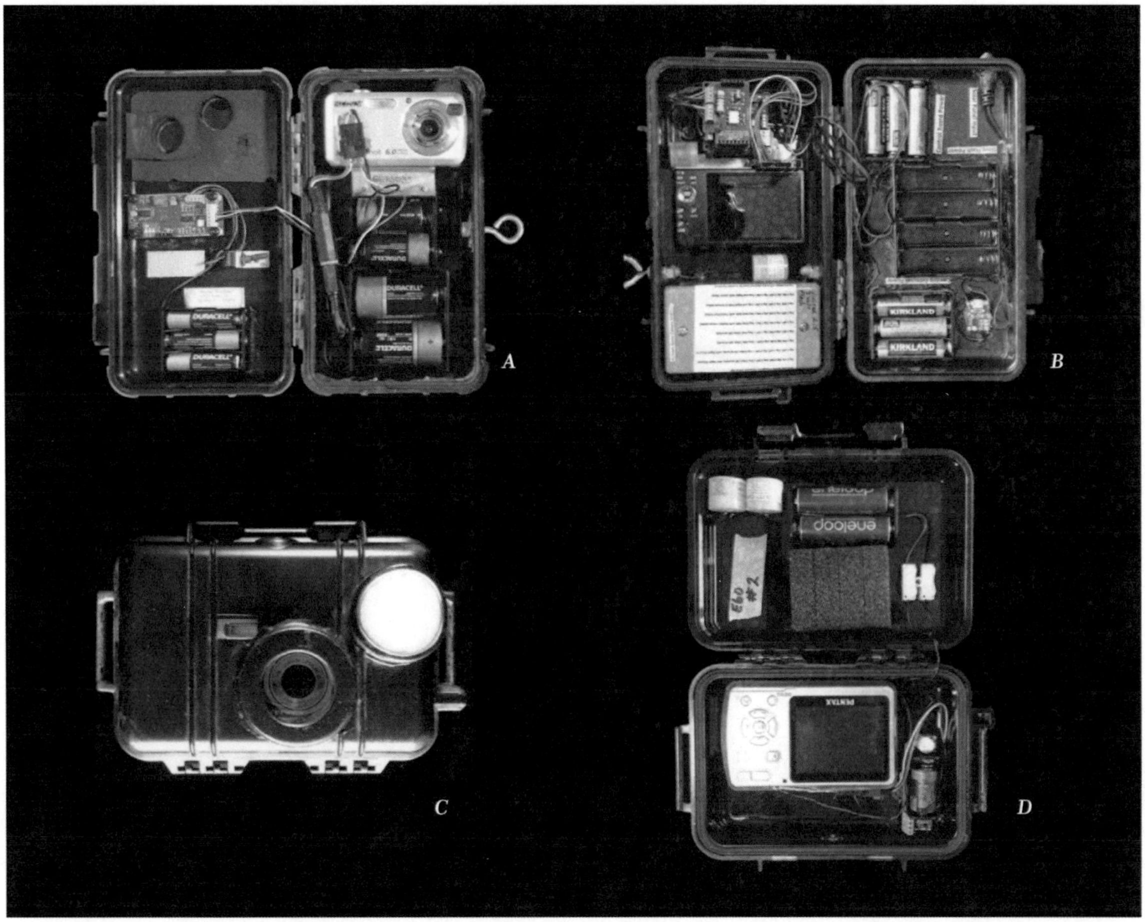

Figure 5.2. Point-and-shoot cameras can be hacked as camera traps. ***A***, This camera trap, which uses a Sony DSC s600 paired with a Pelican 1060 case, was built with external C and D cells for extended use over winter. ***B***, This Panasonic Lumix DMC SZ9 was hacked to take stills and HD day videos. ***C–D***, These photos of a Pentax Optio E60 show the terminals wired during hacking and the finished camera with a pigtail connector.

Many homebrews have been built from point-and-shoot cameras that can autofocus subjects from about 12 inches to infinity. Their wide-angle lenses (approximately 28mm) give almost twice the FOV of most commercial trail cameras and offer more control of settings, such as flash intensity, ISO, and exposure value.

Homebrewers have experimented with various models of point-and-shoot cameras, which they electronically modify or "hack" (Figure 5.2). Models that take **high-definition (HD)** video by day and still photos by night using an external flash were a recent trend but have already fallen out of favor. Other homebrewers prefer DSLR cameras, which can be adapted simply by connecting the controller to the remote shutter release jack (Figure 5.3). Even GoPro and Sony action cameras have been customized as camera traps (Figure 5.4). With patience, modest electronic skills, and help from online discussion groups and tutorials, you can build a quality camera trap. Or you can pay an experienced camera hacker to build one for you.

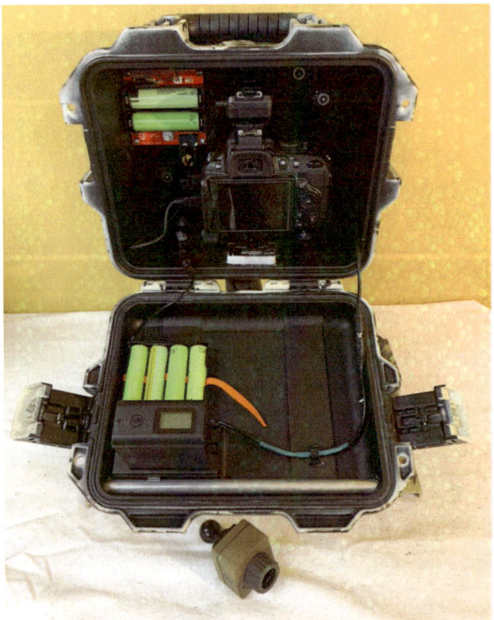

Figure 5.3. A Nikon D5200 DSLR camera wired as a camera trap and housed in a Pelican iM2075 Storm Case. ***Foreground***, A wireless TRLcam motion sensor triggers the camera. ***Upper left***, Two 18650 batteries power the motion sensor. ***Lower left***, The four batteries power the transmitter on the camera's hot shoe, which activates external flashes. ***Lower right***, Beneath these batteries is the NP-F battery that powers the camera.
Courtesy of Daryl Dean

Conclusions

Homebrewed cameras are not as user-friendly as commercial trail cameras. Most commercial trail cameras have fixed-focus lenses and relatively low-resolution image sensors, but their images are interpolated to increase megapixels (see "Picture Resolution"). Homebrewed cameras have autofocus lenses and higher resolution image sensors. Both types of cameras take color video by day; and almost all commercial trail cameras take infrared night video, while few homebrewed cameras can be hacked to do so.

There are also differences in the time and effort needed for camera deployment. Commercial trail cameras are easier to set than homebrewed cameras, especially DSLRs and action cameras when used with external accessories, such as flash units and triggering devices. Visualizing the FOV accurately is difficult with trail cameras that lack LCD viewfinders, but this improves with experience. Older model action cameras like the GoPro also lack viewfinders, but they do not demand precise framing when set for wide-angle shooting.

A *B*

Figure 5.4. Two GoPro Hero2 action cameras hacked as camera traps. Using one or more remote RF triggers is a remedy for this camera's four-second reaction time. **A**, An all-purpose build using a Safari controller and three 18650 batteries for back-up power. Velcro near the lens is used to attach a makeshift lens cover. ***B***, The GoPro "rat cam" is designed for video recording small mammals and ground birds. Three 18650 batteries supply back-up camera power. A 12-volt battery bank powers the 9-watt white LED lamp on a telescoping post.

BATTERY FACTS

Trail cameras differ in battery power, and many desirable features drain this power. Power drain is greater for white flash than for infrared arrays and for video clips versus still pictures. LCD displays are an additional drain. Some commercial trail cameras allow extended battery life with an accessory external battery pack or a solar charger. Battery life also depends on weather conditions: hot and very cold weather reduce it.

Alkaline batteries are relatively inexpensive, have a long shelf life, and last longest in low-drain cameras, but they cannot be recharged. Voltage for fresh batteries is about 1.5 volts. These batteries give reasonable performance in the controllers of homebrew cameras but discharge quickly in digital cameras hacked as homebrews. When in use, the rate of voltage drop is steady. Premium alkaline batteries cost more but last longer in high-drain electronics.

NiMH (Nickel–Metal Hydride) batteries are more costly and have a shorter shelf life than alkaline batteries, but they perform longer in high-drain devices, like most cameras. NiMH batteries are also rechargeable. Since they lose about 1 percent of their charge each day, they should be charged immediately before use. Maximum voltage of NiMH batteries is about 1.2 volts, though they register as high as 1.4 volts immediately after charging. When in use, the voltage declines slowly and then suddenly plunges. Battery capacity is the amount of current a fresh battery can deliver over twenty hours at 68 degrees Fahrenheit. It ranges from 1,200 to 2,900 mAh depending on the brand, and higher capacity batteries last longer.

LSD, or **low-self-discharge batteries**, such as the Sanyo Eneloop, come pre-charged, and shelf life exceeds that of regular NiMH batteries by half. They resemble regular NiMH batteries in other characteristics, but battery capacity (mAh) is less in LSD NiMH batteries.

Energizer Ultimate Lithium (lithium) AA cells have a very long shelf life, are excellent for high-drain devices, and perform well at subfreezing temperatures, but they are not rechargeable. Their the best choice when shooting video or camera trapping in the winter and their working charge is between 1.7 and 1.8 volts. You should replace them when they reach 1.6 volts.

Rechargeable lithium-ion (li-ion) batteries are the proprietary power sources in digital cameras. Lithium-ion cells (18650) resemble large AAs with an operating voltage of 3.7 volts. They vary in capacity and are often used to power homebrewed cameras and LED spotlights. It's a good idea to buy protected cells, which have a circuit to prevent overcharging.

CAMERA TRAP ANATOMY

Modern camera traps, both commercial and homebrewed, have four basic components: camera, circuit board (called the controller), power supply, and housing. The controller integrates the performance of elements such as sensors and flashes. External incandescent flashes or LED arrays are optional elements.

A

B

Figure 5.5. Screenshots from commercial trail camera videos often show motion blur or image stretch, which makes it difficult to study the mechanics of movement. Though blur is visible in single video frames, the action flows smoothly even when edited for slow motion viewing. ***A–B***, Differences in motion blur may be more apparent in clips shot in daylight. This squirrel was apparently moving faster than the deer. (Browning Recon Force Extreme, 60 fps with motion blur reduction)

The **controller** is a printed circuit board with a preprogrammed microcontroller chip (also called a microprocessor). Functionally it is the camera's brain, perceiving the world through eyes called **sensors**. The **ambient light detector** (or **day-night sensor**) is necessary for operating under different light conditions. The PIR sensor (or **motion detector**) senses heat in motion, which is necessary to activate the camera shutter. The controller integrates these inputs with other operations.

The LCD shows the camera's status and settings and may also display stills and movies. Programming buttons allow you to select the camera's modes of operation, and the **walk test** function allows you to adjust the camera's **detection zone** before it is ready to take photos. Just walk in front of the camera at various distances, and the flashing, light-emitting diode (LED) shows you where you—and other warm-blooded animals—can be detected.

AA cells are used to power most commercial trail cameras, and battery status can be monitored on the camera's LCD. Commercial trail cameras commonly have a 12-volt jack for extended power from an external battery, which can be maintained with a solar battery charger. YouTube videos show how to rig your camera for prolonged operation in the field[3] (see "Battery Facts").

Camera trap cases are made of durable plastics and are usually weatherproof or weather-resistant, if not moisture proof. Moisture penetration is often a problem in humid climates, and though recoveries from immersion have been recorded, none are submersible for an extended time. All camera traps can be secured to trees, posts, stakes, or even rocks using lockable cables, straps with cinches, tripod heads, or screws and mounting brackets (see chapter 9). Metal security cases are readily available for commercial trail cams and can be constructed for homebrews.

CRITICAL CAMERA TRAP FUNCTIONS[4]

Trail cameras share a set of basic features and are generally easy to use, but there are functional differences between makes and models. Judicious reading of product reviews can guide you in choosing which camera suits your needs. Some buyer complaints are clearly due to lack of knowledge and experience. You can only learn your camera's strengths and limitations by using it. Camera manufacturers often use terminology that is brand-specific, and I have included some of those terms commonly in use.

Infrared Versus White Flash Cameras

The trail camera industry named the infrared LED array the camera's *flash*. This misnomer conflates the extremely brief bursts of **white light** produced by electronic flash or **strobe light** with continuous illumination produced by traditional photographic floodlights and LED arrays.[5] Incandescent xenon flash can freeze most animal movement because it lasts hundredths to thousandths of a second. Infrared light from a trail camera's LED array is actually continuous light of short duration (a little less than a second), and it is not bright enough for the shutter to freeze quick animal movements in still pictures or videos, even at 60 fps (frames per second).

White and infrared flash illuminate differently at night. Infrared LED light gives monochromatic or grayscale images, and white incandescent light gives color images. Most trail cameras on the market are equipped with infrared LED arrays, and their radiation differs in visibility. **Low-glow** arrays emit wavelengths less than 940 nanometers. Look at the LEDs directly and you will see a dim red light; most mammals and owls show signs of seeing it too. Some animals seem to react to the beam of light even when the LEDs are not visible to people.

No-glow arrays produce wavelengths greater than 940 nanometers. These are invisible to people (unless viewed with a digital camera) and probably can't be seen by most mammals and nocturnal birds. Some no-glow arrays emit a deep purple light visible to people only at close range. Commercial trail cameras featuring black, ghost, or no-glow flash, are often called **stealth cameras.** (Chapter 11 discusses the effects of flash and infrared light on wildlife).

Infrared trail cameras take color images in daylight and infrared images in dim light and darkness. An infrared filter renders accurate color in daylight, but without it, color photos have a pinkish cast. To detect infrared illumination at night, however, the infrared filter must be removed. The **IR cut switch**, or **IR exchanger**, is the mechanism that makes this possible, and it often makes a slight noise as it flips into position.[6] Some newer models of trail camera have separate lenses dedicated for day and night operation and thus do not need cut switches.

MORE ABOUT INFRARED SENSORS

PIR sensors are the most common trigger used in camera traps. These heat-in-motion detectors are sensitive to the infrared band of the energy spectrum. Infrared energy is heat generated by the metabolism of animals, but it also radiates from nonliving features of the environment that are heated by sunlight. A **Fresnel lens** is used to focus the infrared pattern on the sensor. The detection zone is usually cone-shaped with an oval or elliptical cross-section. Up to a point, the detection area increases with greater distances from the lens.

AIR sensors transmit an infrared beam to a receiver. When an animal breaks the beam, the infrared sensor is activated. TrailMaster monitors allow the user to adjust the length of time the beam is broken, which is an added advantage in fine-tuning the device to detect animals of different size and velocity.

C

D

Figure 5.5 *C–D*., Image stretch was only recorded in a single forelimb of the ringtail as it crossed a limb, while the entire body of the airborne jackrabbit was stretched. The camera and settings were identical in both photos and the same as in used in ***A*** and ***B***.

Flash range is a vaunted selling point of commercial trail cameras. Infrared lighting commonly has a maximum range of 50–100 feet, and models with a long-range setting may reach 120 feet. Infrared can capture images and videos of animals at greater distances than white LEDs and xenon flash, which usually have a range of 40 or 50 feet, respectively.

Infrared trail cameras cannot record quick animal movements sharply because they emit light of relatively low intensity, and the shutter speed is longer than necessary to freeze fast-action. Some models, however, offer **blur reduction** or **high-speed** modes (60 fps) to improve clarity. Single frames from videos taken in these settings show less "stretch" of the images but still make acceptable slow-motion clips (Figure 5.5*A-F*).

Trail cameras limit the maximum duration of infrared nighttime video to fifteen or twenty seconds to conserve battery power. Prolonged shooting modes like **IR Smart Video**[7] that override clip length do not apply to night video. If you do not want to miss longer action sequences, set the camera for the shortest photo delay; but be aware that when battery power is low, the camera will cut night video duration to a few seconds. Despite low battery power, daylight videos draw far less current and may take full-length videos for days.

Infrared stills and videos may overexpose subjects at close range. It depends on the camera. Recent models boast **adaptive lighting** or **automatic dimming** to maintain optimal exposure. This is a desirable feature if you want to photograph wildlife at night and at relatively close range.

Homebrewers used to hack point-and-shoot cameras as infrared cameras and used them with electronically triggered external flashes.[8] These cameras took amazingly crisp photos. Eventually, point-and-shoot cameras with extendable lenses replaced the older models, and lens extension made it impossible to add the IR cut switch. Commercial trail cameras have not yet matched the quality of night imagery possible with the older point-and-shoot cameras.

Night videos taken with LED arrays are visually stunning, but routing electrical cables from the camera to LED arrays is an ordeal. The solution is to use wireless LED arrays with a **radio frequency (RF)** transmitter mounted on the camera's **hot shoe** (Figure 5.6). It's an added expense, but it greatly simplifies the work of setting up. It also reduces the risk of a cord-entangled buck scattering your equipment.

E

F

Figure 5.5 *E–F*., The stretched eyeshine of a Great Horned Owl revealed it stooping on a jackrabbit, which escaped the attack and hopped into the frame a few moments later.

PIR Sensor

PIR sensors are the standard motion detector used in commercial trail cameras. There are two elements—a translucent plastic Fresnel lens and the sensor. The lens focuses ambient infrared energy on the sensor. The detection angle depends on the curvature of the lens and on the pattern of concentric rings etched in the plastic, called Fresnel zones. Animals can shutter blank photos if the detection angle is greater than the lens's angle of view. Some camera trappers view this design feature as a deficiency, but those who favor video prefer the camera to be rolling when the animal walks into the FOV.

The sensor is a small **pyroelectric** grid connected to the controller's circuitry, and it works like a concentrated array of temperature sensors. Circuitry detects heat moving across the temperature grid and triggers the camera. Large bodies are detected more easily than small bodies, and movement across the field is more detectable than movement toward the camera. Sensitivity is greater with low background temperatures and is prone to **false triggers** at high temperatures. The optimal range of temperature is 60–80 degrees Fahrenheit. Sensitivity adjustment allows some degree of control over thermal noise produced by warm breezes, rain, and falling snow.[9] PIR sensors have a brief calibration period when the camera is powered, usually followed by the walk test (see "More about Infrared Sensors").

A

B

Figure 5.6. Wireless flash triggers are a welcome alternative to cables when deploying LED flood lights for night video. **A**, The wireless Pixel Pawn transmitter is attached to the camera's hotshoe (seen in the lower right corner of the case) is wired to the camera's controller. ***B***, The LED floodlight case contains the wireless receiver and the 18650 batteries needed to power the receiver and floodlight.

Reaction Time (Trigger Speed, Trigger Delay)

When a trail camera detects a PIR event, there is a delay before the picture is taken. This is the reaction time. Most commercial trail cameras have reaction times of 0.1 to 1.5 seconds. Homebrewed point-and-shoot cameras react slightly slower at approximately one to three seconds. DSLRs shoot in less than a second in manual mode but take considerably longer in autofocus mode. The distance between camera and subject is also a factor because the FOV increases with distance. Slow cameras can photo-capture distant animals because the zone of detection there is wide. At close range, however, running animals may cross the narrow FOV before the camera can react. You can compensate for slow reaction time by aiming the camera at a tangent to the animal's travel route or trail. This gives the camera more time to react while the animal is in the FOV. (See chapter 6, "Triggering the Camera at the Right Time.")

Memory

Trail cameras can store up to 512GB of images on removable SD cards, depending on make and model. The highest capacity is rarely filled unless cameras are left out for months at a time, or animal traffic is high. Scavenging birds at large carcasses can fill a card quickly. Hot windy weather alone can fill a 32GB memory card with blank video clips in a few days. Checking time stamps or your EXIF data will tell you when the wind starts and ends. If your camera can schedule shooting time, you will save memory and thwart frustration.

Recovery Time

Cameras need time to record images and video clips to memory. Recovery time tends to be brief—less than a second for still pictures and longer (up to a minute) for videos in commercial trail cameras. Instant recovery time is possible in point-and-shoot and DSLR cameras that have more **RAM** or buffer space than trail cameras.

OPERATIONAL SETTINGS OF COMMERCIAL TRAIL CAMERAS

The LCD and programming buttons allow viewing and selecting of camera functions, and camera operations in Wi-Fi cameras can be remotely set with a cell phone. Be aware that some manufacturers substitute invented terms like **field scan** for commonly used ones like time-lapse. This book uses conventional terminology and some synonyms commonly used in the trail camera trade.

Photographic or Operation Mode

Still pictures, often called trail mode, and video are standard in most commercial trail cameras, and time-lapse is an added option in many models. Some cameras can run time-lapse mode and trail mode simultaneously. This is often marketed as field scan mode. Others offer a hybrid capture mode that takes a still photo with each video clip. A few trail cameras are dedicated exclusively to still or time-lapse photos.

Picture Interval (Delay) and Video Length

The interval between pictures (1 second to an hour or more) is a "**time-out**" during which the motion sensor is unresponsive to PIR events. Short delays allow the camera to trigger successive photos or video clips of animals lingering in the detection zone. They increase your chance of multiple exposures and views, and of recording something unusual. Long delays conserve memory by limiting redundant photos triggered by "camera hogs." There's a price to be paid for short picture intervals, and that is the time needed to review images and clips of commonplace species, but sometimes even those can give you a pleasant surprise.

Maybe you want short picture delays to get a choice of pictures. Many cameras can take multiple shots per triggering event. Select the desired number of photos from the menu, generally two to ten, and the camera will shutter a sequence over a period of a few seconds. **Rapid fire**, or **burst mode**, fires a quick volley of images with intervals of less than a second. Image bursts have smaller files than single images. Use these shooting modes with a specific goal in mind. When shooting five pictures per trigger, you will take five hundred photos if animals trigger your camera a hundred times, in addition to other exposures from false triggers.

Time-lapse allows photos at fixed, preset intervals and yields an abundance of still images, which can be made into accelerated videos of normally slow phenomena. Video length can be set from about five seconds to two minutes, but IR Smart Video allows the camera to shoot videos up to five minutes as long as the motion sensor detects movement.[10] You don't want to use smart video in "false trigger country." Your camera will shoot long clips wherever the wind blows and the vegetation dances, and you will be the poor sod who has to view it.

Picture Resolution

Choice of image size in digital cameras allows you to take more low-resolution or fewer high-resolution images. Small images are appropriate when using smaller gigabyte memory cards or when images are intended for the internet. Current models take photos ranging from 3 to 20 megapixels (MP), but trail camera makers use a digital sleight of hand called interpolation to enlarge the true size of the images.[11] In cameras with this feature, the microprocessor automatically enlarges the image and fills the gaps between the original pixels with new, color-matched pixels. An image from a 2MP sensor can be doubled or tripled, and thus a larger megapixel image can be produced with a less expensive and smaller image sensor. When researching your choice of trail camera, it's important to read reviews to find the few models that don't interpolate. If high-resolution images are your goal, you will prefer to use a homebrewed point-and-shoot camera or a DSLR.

Shooting Time

The ability to select the time frame of operation (day, night, twenty-four hours, or selected hours) differs between cameras. Many models in the past gave you no choice—the camera operated around the clock. Others could be set for daylight or night operation. More recent models offer two options: 24-hour operation or a capture timer that allows selection of specific start and stop times. This is a welcome feature if you set cameras where extreme heat or wind triggers endless **blank images**. You can pick an operation time during optimal lighting conditions or when your subject is known to be active.

PIR Sensitivity

PIR sensors differ in detection angle, sensitivity, and tendency to trigger to ambient (nonanimal) infrared noise. Commercial trail cameras allow automatic sensitivity selection or manual selection of several levels of sensitivity. Some manufacturers have reduced the PIR sensor's tendency to record false images by using two Fresnel lenses or by dividing the Fresnel into sectors.

OTHER FEATURES

The main trail camera features available at the time of writing are outlined below, but the industry will certainly continue to modify and add new features in the future.

Data Strips

Trail cameras that record moon phase, temperature, barometric pressure, and GPS coordinates can be programmed to print the data strips on the image or video clip. Data strips can be a distraction in movie productions, but they can be easily removed by cropping with photo editing software. All digital cameras record date, time, and exposure variables in EXIF data files that are readable with various programs and applications. Still images and video clips differ in the kinds of data recorded, but both record date and time. (Chapter 14 discusses how to use this information.)

Image Viewing

The LCD image viewer is not a feature of all trail cameras, but it offers an easy way to preview the photo-catch in the field and is a great aid to framing scenes. Portable viewers offer another way to see your pictures, or you can upload images to a laptop or tablet. Ultimately, however, you will download pictures and video clips to your home computer using a digital **USB** cable or a card reader. You can also view them on a television screen using the audiovisual out cable supplied with the camera.

Status Information

The photo counter on the LCD tells you at a glance how many photos or clips were taken, as well as battery status and shooting mode. Some models display unused video time remaining on the SD card, rather than the number of clips remaining.

Attachment and Security

Trail cameras come with a standard 1/4-inch-20 threaded tripod socket for use with screw mounts. Mounting slots and a strap are also supplied for attachment to tree trunks, and other mounting devices are available or can be made. Lockable boxes and steel cables are the most common means of securing cameras against theft. Password security is another feature that will frustrate if not deter a thief. (Chapters 9 and 10 discuss security and accessories in detail.)

Wireless Photo Upload

One of the latest advances in trail camera design is wireless transmission of images and video clips to cell phones or email accounts. Wireless trail cameras automatically transmit files via Wi-Fi or cellular network and alert you when battery power is low. Convenience comes at a cost. Service providers require a separate phone number or line for each device, though there are several payment options. Another innovation is short-distance wireless transmission to a black box receiver, where images are recorded to an SD card. Short-distance Wi-Fi transmission to local networks is becoming more common in many trail cameras. A thief might steal the camera and fail to see the receiver, which can be placed up to 150 feet away.

HOMEBREWED CAMERA TRAPS

Homebrewed camera traps offer more customization and design options than commercial trail cameras. You can buy used and new homebrews on hunting and camera trapping forums and from professional builders. Lower-end point-and-shoot cameras may cost less than $100, but with other components a homebrew can cost $150–$250 or considerably more.

Making your own camera trap from parts can be a rewarding learning experience, even for those who are not natural-born tinkerers. The happy homebrew user has a basic understanding

SOME CAMERA TRAPPING FORUMS

Forums dedicated to hunting and camera trapping have decreased in number, but those listed here are helpful resources for new and seasoned camera trappers.

- *Camtrapper*, DIY Camera Trap and Wildlife Surveillance Forum: https://camtrapper.com/
- *Camera Trapper:* https://cameratrapper.com/
- *Wildlabs.net*, The Conservation Technology Network: https://wildlabs.net/discussions
- *Jesse's Hunting & Outdoors:* https://jesseshunting.com/forums/game-trail-cameras-camcorders.273/

of how their camera works, and the frustrated user is usually one who is afraid to tinker. Help is available in a community of builders in the United States and Canada, and there is a growing community in other countries.

GETTING STARTED

Online information about camera trap construction is indispensable to new camera trap hackers. These sources include searchable do-it-yourself postings and YouTube videos, but the most interactive and diverse resources are found on a small number of forums where communities of camera trap hobbyists freely share knowledge and experience (see "Some Camera Trapping Forums"). Registration is free and allows direct participation in ongoing dialogues about camera trap construction.

Forum members can provide information about cameras suitable for hacking. Research and development (R & D) has been the personal initiative of a few gifted hackers who solicited members for interest in recent models, and some members even made donations to purchase the camera and supplies. The test hack results in an assessment of the camera's suitability for modification. Successful R & D projects have given rise to instructional materials such as YouTube videos and PDF files of step-by-step building protocols (see "The Search for the Ideal Homebrew Camera").

HOMEBREW BUILDING OPTIONS

There are several degrees of difficulty in building camera traps. It is relatively simple when using a prefabricated kit, more challenging when building from parts. Let's consider the options.

Option 1: Hack Camera, Assemble A Kit

The easiest way to build your first homebrew is to buy a kit and a digital point-and-shoot camera that is commonly hacked.

PHASE 1: RESEARCH

You'll need research and preparation to start, so begin by familiarizing yourself with the building components and the process of hacking the camera. Here is a step-by-step scenario.

1. Browse trail camera forums; explore all the sections and read the posts. Register in one or more forums and keep reading. Photos posted by members will reinforce your interest.

2. Visit web pages of various suppliers of camera trap parts (see "Some Camera Trap Building Supplies and Services"). Read the materials, including the instruction manuals about the control board and camera modification. Familiarize yourself with the building process. Email questions to the vendor or post them on a forum.

3. Decide whether you want to build your own camera or buy a new or used homebrewed camera through a forum or from a vendor.

4. If you decide to buy a used camera trap from a forum member, caveat emptor. Get the seller's assurance in writing that the camera is fully operational. Forum managers are not responsible for transactions that go bad.

PHASE 2: PURCHASE A CAMERA AND ASSEMBLE YOUR SUPPLIES

If you decide to build a camera trap, buy the camera and components and then assemble tools and supplies (see "Basic Camera Hacker's Kit").

1. Buy a camera that is commonly hacked. You may buy new, refurbished, or used cameras from retail stores, Amazon, eBay, or sellers on camera trap forums.

2. Buy the other components: the case, controller, glass circles, Fresnel lens, battery holders, etc. First-time builders used to buy a prefabricated kit, which included a predrilled case fitted for a specific model of camera, but nowadays kits are hard to find. Vendors or forum users may be willing to drill the case for you.

3. Purchase the camera parts: controller, Fresnel lens and mounts, case, battery holders, etc. Get the supplier's advice if there is more than one option available for a part.

PHASE 3: CONSTRUCTION

You are about to embark on a brave mission, so do your homework and know what to expect. Once you open the camera, its warranty is void.

1. Download the camera modification protocol and plan the steps in the hacking process. Study and take notes from YouTube videos, if available.

2. If soldering is new to you, refer to a book or the internet to learn how to tin wires and contacts. Be sure you are proficient before you start on the camera.

BASIC CAMERA HACKER'S KIT

TOOLS

- Multimeter
- Cordless drill or Dremel tool with small bits
- Wire stripper and cutter
- Wire cutter
- Fine-tipped soldering iron
- Jeweler's screwdriver set
- Disposable lighter
- Magnifying glass or lamp
- Locking clamp (hemostat) (optional)
- Fly tying vise (optional)

SUPPLIES

- Rosin-core solder (60/40) (0.032" diameter)
- Soldering paste flux
- Assorted insulated wire (26–28 gauge) of different colors
- Shrink insulation (various fine sizes)
- Electrician's tape

3. Be methodical. Use a clean working surface and put the screws in a container. If there are a lot of screws of different sizes, tape them to a piece of paper and label them.
4. Be careful. Cover flash contacts with electrician's tape. Getting a shock from a hot contact won't kill you, but you won't forget it.
5. Work patiently. Don't panic if you think you made a mistake.

If all this sounds intimidating, there is another way to hack your own camera. Find a local mentor through a trail camera forum. Members of some forums have convened periodic workshops or "cam fests," where seasoned camera trap hackers guide newcomers through the steps. Workshops can be highly educational even if you have already hacked a camera.

Option 2: Hack Camera, Build from Parts

This option is for the do-it-yourselfer who prefers to economize and shop for the best deals on each component. Drilling cases for flash, lens, and sensor is an exacting task that requires precision measurement, power tools, steady hands, and patience. It helps to make jigs for the layout of the holes because plastic cases have curved surfaces and are difficult to mark accurately for drilling. Accomplished do-it-yourselfers have developed the most compact homebrew designs and layouts. Members of camera trapping forums can provide helpful advice.

Option 3: Hack Camera, Program Microcontroller

A number of camera trap builders program their own microcontrollers (or microprocessors) located near the center of the controller card. All microcontrollers require programming for their operation and compatibility with different camera models. Suppliers preprogram their microcontrollers for specific cameras, which is why you must specify the camera when ordering a controller board.

However, you may build and program microcontrollers, such as the **Arduino**, using **firmware** and the **PICAXE**-08M2 chip. You can get programs from vendors[12] or learn to write programs on your home computer and copy them to the microcontroller using a special cable and a circuit board with a chip receptacle. Members of camera trap forums have also posted programs they have written for different cameras.

Some camera trap enthusiasts have made their own controllers from electronic parts and have advised about construction and programming on camera trap forums. Such homemade controllers have the advantage of versatility and can be wired and programmed for specific needs.

Option 4: Adapting DSLRS as Camera Traps

DSLR cameras are readily adaptable as camera traps and are more versatile than point-and-shoot cameras. Hacking is relatively easy because the electronic remote control terminal can be connected to a controller without opening the camera. DSLR camera buffs often use off-camera motion sensors, which offer possibilities for staging photographs not possible with PIR controllers. Excellent step-by-step instructions for a variety of DSLR camera traps are available at several online sources.[13]

For those interested in high-speed photography, several triggering devices are available for use with DSLRs. The controller is compatible with a variety of sensors that photographically freeze the motion of flying birds, bats, and insects, but limited battery power makes them most suitable for sets of short duration.

In general the camera, trigger accessories, and electronic flash units are not self-contained like commercial trail cameras and homebrewed camera traps.

Advanced camera traps can be more costly, more complicated to use, and more time-consuming to set up and maintain, but they offer photographic possibilities not available in homebrewed point-and-shoot cameras. (See chapter 6 on deploying DSLR camera traps.)

A

B

Figure 5.7. Photos taken with a Canon A630 camera hacked with CHDK:
A, a deer mouse *(Peromyscus maniculatus)* and ***B***, a dusky-footed woodrat *(Neotoma fuscipes)*.

Option 5: Adapting Canon Point-and-Shoot Cameras as Camera Traps

Many digital cameras manufactured by Canon can be programmed with software scripts to become time-lapse cameras and camera traps. The microcontrollers in these cameras have many operational features not mentioned in the owner's manual, but these operations can be activated using remote programs. Developers from various countries have written programs and tools known as the **Canon Hardware Development Kit (CHDK)** to make the advanced features accessible to users.

The software scripts and programs are specific to over seventy camera models but can be easily downloaded, copied to the SD memory card, and loaded and run on the camera without affecting the warranty or changing the camera. Canon cameras also commonly have a DC-in receptacle that enables an accessory power source. Among the functions of interest to camera trappers are ultra-high shutter speeds, **RAW** images, time-lapse, motion detection based on pixel changes, and remote triggering with PIR sensor using the mini-USB port. **Pixel change detection** works quite well for close-up flash portraits of rodents (Figure 5.7*A–D*).

C

D

Figure 5.7 *C*. The Canon A630 can use its own pixel change detection only for daylight shooting, or it can use a PIR sensor *(above the camera lens)* for shooting around the clock. External flashes are needed for uniform lighting in this particular build because the snorkel casts a shadow when the camera flash is the sole source of lighting.

Figure 5.7*D*. An overnight set using a "naked" Canon A630 triggered with an overhead PIR sensor. Two external flashes equipped with photoelectric triggers supplement light from the camera's flash.

THE SEARCH FOR THE IDEAL HOMEBREW CAMERA

The homebrewer's dilemma is to find the ideal camera for a camera trap. Digital point-and-shoot cameras are not designed with the camera trapper in mind and often differ in operation and capabilities.

The ideal camera has a quick shutter, strong flash, good optics, and high-resolution. It is easy to open and modify, which means the circuits have large contacts for easy soldering. Use of nonproprietary batteries, like AA, is also a plus because backup power of compatible voltage is easier to install. Seldom will you find a camera that offers all of the ideal attributes in a single package, and most of the less appealing features can be worked around.

Theoretically, camera manufacturers could produce a camera that is particularly suited for remote use, but the main developmental themes of point-and-shoot cameras are smaller size, larger LCDs, higher resolution, and zoom lenses with greater ratios of focal length. Now smartphones lead point-and-shoot cameras in the arena of innovation.

Some advancements are welcome improvements. Others, such as extendable zoom lenses, are not usable in most homebrews and complicate the building process. But they can be overcome with the use of a snorkel, which fits on the front of the build case and allows the lens to extend unobstructed. Others, such as touch-screen menus, are of little use.

Option 6: Using CCTV Cameras as Camera Traps

Closed circuit television, or **CCTV**, is an excellent way to view wildlife in real time and can also be used to monitor and record videos remotely (Figure 5.8). The technology has been steadily improving, and many cameras are available across the price range. CCTV is divided into two technologies: composite and IP (internet protocol). Composite CCTV is the original closed circuit TV monitoring method. When this technology entered the market, the cameras were limited in resolution and storage was difficult. **IP cameras** made worldwide video transmission possible through the internet, and high megapixel cameras have become an affordable and excellent tool for monitoring wildlife. Basic CCTV components include a relatively small camera, a recording method, and a power source, usually POE (power over ethernet) or 12-volt DC.

Several features distinguish CCTV cameras from trail cameras. Generally CCTV cameras sense movement using software to detect pixel change, use infrared LEDs for night illumination, and transmit images continually to a TV or computer monitor using cables or wireless transmission. Many IP cameras store images and video in the camera on an SD card or microSD card. Those images can then be acquired remotely by logging into the camera and downloading the files. This method works especially well when the network bandwidth is not sufficient to permit streaming of high-quality video. This also eliminates the need to have the camera connected to a computer or network storage device.

A camera trap using an IP camera with internal storage only needs the camera, up to 300 feet of **Cat 5 or Cat 6 cable**, and a power source, which can be provided by a solar panel, a charge regulator, and a battery. Setup is done with a laptop. When the site is visited to download the images or video, the actual camera site does not need to be disturbed.

IP cameras have a built-in web server used to configure the camera features and monitor the video. To configure the motion detection option, the user can block out zones where there is unwanted nonanimal movement, like vegetation, that could trigger the camera and fill the memory card. A big advantage to IP cameras is the ability to always have the last five seconds of video continually stored in a buffer. When motion is detected, the camera can actually go back in time by five seconds and start recording from there. So, rather than seeing the animal at the start of the video, you can actually see the animal enter the scene.

Figure 5.8. An IP camera (Dahua 2MP IR Bullet Network Camera) with alligator cables for connecting to a 12-volt SLA or car battery. The camera base is attached to a cap that fits over the end of ABS pipe container.

Real-time viewing of feeders and nest boxes is the simplest application (Figure 5.9). If you already have a CCTV home security system, this setup only requires dedicating one or more cameras to a feeding area, salt lick, trail, or nest box within wireless or wired transmission range of home. Remote recording is more complicated to set up using cables, but wireless is an alternative that is workable within 300 feet or so of the house. Transmission over longer distances is also possible, or a portable self-contained unit can be used. Susan Young's book *CCTV for Wildlife Monitoring* is an excellent introduction to all the possibilities.

Figure 5.9. My neighbor wired this CCTV bird-nest box to his house. He could view oak titmice rearing their young in his TV room.

SOME CAMERA TRAP BUILDING SUPPLIES AND SERVICES

- **Getting Started with Arduino Products:** https://www.arduino.cc/ (circuit boards that can be programmed as camera controllers)
- **Camera Axe:** http://www.cameraaxe.com/wp/ (high-speed DSLR controllers and accessories)
- **Camtraption Remote and Camera Trap Photography:** https://www.camtraptions.com/camera-traps/ (DSLR accessories, motion sensors, housings, flash triggers, remote control buggy, and supplies)
- **Cognisys, Capture the Hidden World:** https://cognisys-inc.com/learn-how-to/camera_trapping/ (camera trap cases, sensors, lighting, and accessories)
- **Herd 360, Build Your Own Trail Camera Battery Box Parts and Kits:** https://www.herd360.com/collections/diy-external-power-products/ (cases for cameras and external batteries, chew proof power cables, solar panel controllers and mounts)
- **Snapshot Sniper:** http://www.snapshotsniper.com/ (DIY game trail camera components)
- **TRLcam, Dream, Design, Build:** https://www.trlcam.com/ (custom design and construction of DSLR and IP security camera traps, time-lapse cameras, sensors, and accessories)

CHAPTER 6

Camera Trap Sets: Concepts and Methods

Setting a trail camera can be simple or painstaking, depending on your goals and equipment. Choice of camera placement, perspective, and composition depends on the target species (birds or mammals, large or small) and shooting mode (stills, videos, time-lapse). Camera trapping skills improve with experience. Disappointments decrease when you understand camera settings and limitations, can anticipate behavior of animals on the set, and are aware of Mother Nature's whimsical tricks. Scouting prospective areas benefits from planning and preparation, mapping and GPS skills, and the ability to recognize habitats and read animal sign. Using off-camera lighting and sensors are advanced methods for the use of homebrewed DSLR and point-and-shoot cameras.

A camera trap set is simply a remote camera deployed to record wildlife in a particular location (see "Why Sets?"). It is nothing more than framing a scene to photo-capture expected or hoped for wildlife, but it requires more than just strapping your camera to a tree and turning it on. The focal point of camera trap photos is always wildlife, but the background scene can range from a small patch of nondescript land to a panoramic landscape and distinctive habitat (Figure 6.1*A*–*B*). When you view the scene in the camera's LCD you can see the "stage" where animals will make appearances. If your camera lacks an LCD you have to envision the scene, but by shooting a few test photos you will see its boundaries and details. Animals can appear near and far when the stage encompasses a large area, and their proximity to the camera can be constrained when the stage is small. If you want to show the animal's diagnostic features—head, feet, and tail—you should probably frame a relatively small area. You can create many perspectives of any scene by adjusting your camera's height, angle, and distance to the anticipated subject. You may not be satisfied with

WHY SETS?

The mountain men of the North American fur trade used the word *set* for various techniques of setting traps. There were beaver sets and bear sets, and a *set gun* was a rifle with a trip wire left in the woods. When George Shiras first used the term *set camera*, he acknowledged the automatic camera's similarity to trap sets and set guns, which were all triggered by unsuspecting animals. The current generation of camera trappers have found it a handy term.

TIME-LAPSE SOURCES

- **PhotoPills**: https://www.photopills.com/calculators/timelapse
- **Nature TTL**: https://www.naturettl.com/how-to-create-a-simple-time-lapse/
- **KQED**: https://www.kqed.org/mindshift/17102/slowing-down-time-with-time-lapse-photography/

the resulting photos or video clips, but you will soon see where the main characters usually come and go, and sometimes pause. With that information you can adjust your camera's position to frame the scene more suitably for the next round.

Camera trappers want good portraits and movies of animals doing interesting things. For that you have to tap into your skills as an artist, photographer, and naturalist. The artist seeks perspectives that enhance composition and visual appeal. The photographer makes use of equipment to full advantage, and the naturalist knows that every species can tell its own story if you know when and where to look. Wild animals, however, are unique individuals and free agents, and sometimes they defeat the camera trapper's best-laid schemes. Studying your images and video clips will improve your camera skills, teach you about your subjects, and guide you to improve your sets. In the following sections, I introduce the usual considerations in making sets with respect to commercial trail cameras and homebrewed camera traps.

SHOOTING MODE

Camera trapping goals generally determine your choice of shooting mode—stills, video, or time-lapse. As mentioned above, still photos can portray animals in intimate detail when tightly framed and can cast them in an ecological context in larger perspectives. Wide-angle lenses allow both, presenting a subject etched in fine detail against a stunning environmental backdrop. Like the pioneers of camera trapping, today's DSLR users master wild-animal portraiture through the artful use of lighting and composition and carefully stage their sets to capture the subject as the key element in the picture. Cameras with quick trigger speeds record multiple images of animals that linger within the set and increase the chance of getting a suitable portrait. Stills are also relatively easy to color-correct, store, and share.

Figure 6.1. Two perspectives of a mountain lion: ***A***, on a trail in a nondescript arid habitat, shot with a Sony s600 in San Luis Obispo County, California; and ***B***, as a component in a mountain landscape, shot with a Moultrie M80 Trail Camera in Vancouver Island, British Columbia.
Courtesy of Gary Schroyen

Video buffs prefer to see animals in action. Most clips of animals strolling in and out of the frame fall short of the goal, but locations where animals linger, like water holes and feeding areas, often reveal interesting or even rarely seen behavior. Low-quality video from early trail cameras discouraged would-be nature videographers, but now HD video cameras make high-resolution movies possible. Making even short wildlife movies often means work. It takes time to accumulate clips with behavioral variety to craft a story, and processing and managing video demands more time and computer memory than stills do.[1] But video camera trappers are catching up with action camera enthusiasts that edit recreational clips into quick-moving vignettes and longer sketches.

Time-lapse photography is unique for rendering gradual natural phenomena as dramatically accelerated slices of time. Time-lapse is easy to use with commercial trail cameras. The feature was designed mainly for monitoring animal activity at artificial feeders, but it also makes it possible to accelerate gradually changing weather, tides, sunrises, sunsets, lunar eclipses, plant growth, and other biological phenomena. Taking flawless time-lapse videos requires planning and patience, and practitioners of this genre usually use DSLR and point-and-shoot cameras (see "Time-Lapse Sources").

I first used time-lapse during a trip to Myanmar, where I was intrigued with the large mud castles built by termites. How long would it take these insects to repair damage to their impressive structures? I set the camera at 8:00 p.m., broke off a foot-long pinnacle of hardened mud, and found the castle as good as new the next morning. The film showed that the termites had finished the repair job by 1:00 a.m. I had answered my question and captured footage of an event rarely seen in nature. Nest construction by other social insects (mainly bees and wasps) and birds is a fascinating subject for time-lapse studies but will tempt only the diligent, bold, and hopefully wise.

Many camera trappers find occasion to use still and video cameras in tandem. Some want still images to use as inserts in an edited video stream. Others want to know how animals move through the set and react to the still camera, the flash, or particular props. I used a **watcher camera** to see how bears reacted to an electronic noisemaker intended to protect the other camera from meddling. I learned that my invention was useless.[2] Two or three tactically placed cameras can also capture complementary perspectives that can enhance a movie's visual effect.

A GENERAL PROTOCOL

Setting a camera should be a satisfying experience, but haste and poor planning can ruin the fun. Here are some considerations for developing a routine.

Be Prepared

If your camera is new, study its operation. Then, set the clock and install fresh batteries and an SD card. Format the card with the camera. In the future, you will want to clean the lens and infrared array every time you check the camera. Decide on the various other settings before your outing. You can always change settings in the field, if necessary, but if you don't know the settings menu you may be in for a bout of self-loathing.

Find a Location

Where you set a camera depends on what you are looking for. Suppose you just want to know what species live in the area. If you are a beginner looking for areas near home, set the camera where people won't find it. **Ecotones**, the border areas between two or more habitats, are a good choice because they attract wildlife. Other kinds of habitats can often be found within a hundred miles of home. Let the camera be your teacher. Trail cameras photograph almost every warm-blooded creature that passes, but the chance of photographing many species depends on location. If your objective is a particular species, set your camera where you expect it to live. If you aren't sure, read about its haunts and habits in a field guide or on the internet. (We'll explore this topic in the next chapter.)

Select a Scene

Mammals do not use a habitat randomly or visit all the features in their home range equally, so the chance of your target species encountering your camera is not the same in all areas. Look for signs of wildlife or habitat features that may be used by wildlife. Use of attractants can increase the chances of encounter (see chapter 7), but here we consider sets without such aids.

Assess Lighting Conditions

Changing light drastically modifies a scene throughout the day. The soft light of morning and late afternoon is often interrupted by the glaring sun of midday or punctuated with dark shadows. When sunlight strikes a PIR sensor, false triggers waste power and fill the SD card with useless photos. The angle of the sun and incident light are important considerations for daytime sets. Shaded and north-pointing cameras are less prone to false triggers. The best solution for troublesome lighting in an otherwise splendid setting is to shoot in night mode or schedule the camera to shoot during favorable hours.

Frame the Picture

The location of trees or saplings may dictate where to mount your camera, but a wooden stake, metal post, or small cairn of rock may offer a better vantage point. A camera at the level of the subject usually gives the most intimate perspective. Study the FOV and adjust the camera's position for a view unobstructed by vegetation. Remove obtrusive sticks and vegetation that can cause false triggers in breezy weather. Anticipate **IR glare** from nearby objects. If your camera lacks an LCD viewfinder, go back and forth between your camera and the scene and sight the camera from both directions. Shoot test shots and view them on a laptop. Don't forget to turn on your GPS and record the waypoint, especially if you are in terra incognita.

Apply Attractants

Attractants are optional. Always apply scent after the camera is set. This will prevent inadvertent contamination of your camera and the possibility of it becoming a scent marking target. (Chapter 8 covers this topic in detail.)

Deploy the Camera

Review the camera settings one last time, and make sure a formatted SD card is in place. Record your notes and the GPS waypoint, activate the camera, and conduct the walk test.

DEVELOPING SKILLS AND KNOWLEDGE

Experienced camera trappers understand their cameras' technical abilities and limitations, are perceptive naturalists, and are comfortable in the out-of-doors. In other words, skills and knowledge make camera trapping easy and satisfying.

Know Your Camera

Manufacturers of commercial trail cameras know that most people like things to be simple, and they have made their products easy to use. Their cameras are fully automatic, their menus easily learned, but every make and model acts differently. So, test your camera in the backyard to see if performance matches specifications. Trigger photos and videos as you strut before the camera, or parade your dog about on leash, then view the images on the computer. What are the maximal and minimal distances for optimal lighting at night? In video mode at night, does the camera adjust exposure at close range? What's the best distance for optimal focus? How wide is the FOV at different distances? What is the sensor's zone of detection? This exercise can be enlightening.

Homebrewed cameras have more complicated settings than commercial trail cameras. Animals can trigger photos at any time in changing lighting conditions, so homebrew users optimize their settings for day *and* night, or tailor settings specifically when shooting day *or* night photos. Some DSLR and point-and-shoot cameras, however, can automatically adjust settings when filming in changing light.

Point-and-shoot and DSLR homebrews differ from commercial trail cameras in how they focus images in light and darkness. In daylight, an autofocus sensor measures pixel contrast and focuses the image by maximizing contrast between adjacent pixels. On visually uniform surfaces like sandstone, the sensor can't always meter differences in pixel luminance. Point-and-shoot and DSLR cameras offer several autofocus (AF) modes. Multi-point and center-weighted modes average many

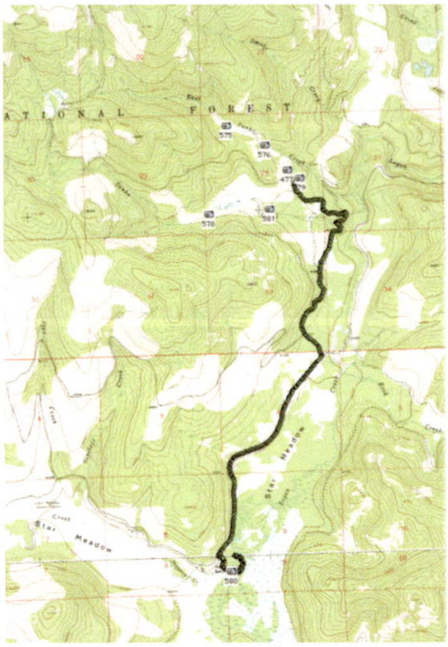
A

Figure 6.2. Two of many GPS tools for recording and displaying camera trapping information. Users can record location notes in both applications. **A**, In MacGPS Pro navigation software, the information is recorded on a United States Geological Survey topographic map, though other maps can also be used. The GPS track shows our route to the site in Montana's Flathead National Forest, and the camera icons mark the locations of our sets.

readings within the scene. They focus subjects within the depth of field and usually give best results. Macro mode allows close-ups of small subjects close to the camera. In darkness, these same cameras adjust focus by measuring the distance between sensor and subject using ultrasound or an infrared beam. DSLR enthusiasts prefer to manually focus a specific area within the frame (see "Advanced Methods: DSLR Cameras with Off-Camera Flash").

Exposure metering also uses different technology in daylight and darkness. Most digital cameras offer several ways to meter exposure in daylight: matrix, center-weighted, or spot. An average exposure is computed from larger views, but an averaged exposure produces a high-contrast image in dappled light or in areas with heavy shadows. In such conditions, using forced flash gives a better image.

Controlling flash intensity for night images is relatively easy if the flash has **through-the-lens (TTL) metering**. The camera's computer measures incoming light from pre-flashes and instantly adjusts the settings for proper exposure. Manual flash requires adjustments of aperture, ISO, flash intensity, or distance between flash and subject.

Homebrewed point-and-shoot cameras may overexpose nearby vegetation and objects at night. The easiest fix is to reframe the scene or remove offending vegetation and objects. If a camera has a weak flash, experiment with higher ISO values or use a slave flash as an accessory. Neither automatic nor manual settings guarantee optimal exposures in all lighting situations. In low lighting, automatic mode uses slow shutter speeds and gives blurred images of moving animals. Manual mode can prevent this when the camera is set for forced flash and a small aperture.

Scouting

Scouting is a logical way to find promising sites to set cameras. Wilderness navigation doesn't come naturally to everyone, which is why topographic maps are invaluable when preparing outings.[3] Maps provide a cognitive framework of the landscape and reveal special features, like springs, marshy areas, and forgotten mines that are often hidden by topography and vegetation. Aerial photos

B

Figure 6.2 *B*. The Google Earth database casts tracks and waypoints (or placemarks) on maps or aerial photographs. Noted information is viewable in pop-up windows.

on Google Earth (https://earth.google.com/) often reveal recent changes in plant cover resulting from fires, floods, and new roads that are not found in older topographic maps. It's disappointing to look at a map after a trip and discover that you missed some intriguing landform beyond the trees. Do your map work ahead of time. It will whet your curiosity.

Plan a general route for a leisurely outing. Your temperament and available time will influence your tactics. Eager camera trappers set a few cameras on the first visit. Others want to explore the area at leisure, take photos and waypoints of potential sites, and return when they are prepared to make specific sets.

Take a compass and a map, or a GPS unit with topographic coverage of the area, and don't forget your watch, camera, water, and food. It is always safer to travel with at least one other person in case of an emergency and to tell someone at home your destination and plans. Remember that cell phone service may not be available where you are going.

Record your movements in unfamiliar landscapes with GPS in tracking mode; the track will also guide you back to your car. You may also want to record potential camera trap sets and points of interest with a camera, GPS, or cell phone. At home, use mapping software to upload tracks and waypoints to a topographic map or Google Earth for study and future reference (Figure 6.2*A–B*).

Recces generally begin on dirt roads, firebreaks, trailheads, or animal trails, but natural features will lure you away from the trail (Figure 6.3*A–G*). Take your time and let your senses lead the way. Hone your skills as a naturalist. Photograph animal sign and unfamiliar plants and animals to identify when you get home. Find a vantage point, sit down, and take in the scenery. The journey is as important as the destination.

Reading Animal Sign

Some naturalists have uncanny abilities to find wildlife. Accomplished birders and **herpers**[5] are admired (or sometimes scorned) because they always see more than their peers. They weren't born that way. Anyone with the motivation can develop skills as a naturalist.[6]

Reading animal sign is one of these skills, and one way to see nature is through your dog's nose. Canines have helped us perceive the world for millennia. They know the world largely by scent and

A

Figure 6.3A. On July 11, we were lured to an oasis—a patch of green on the tawny plain. Six weeks later, we viewed 357 photos of barn owls (*Tyto alba*) in their playground.

Figure 6.3*B–C*. To our delight, photos revealed the owls curtsying, cake walking, and sparring with seed heads of grass. ***D*,** On September 30, a feral pig toppled the camera. ***E*,** The "barnies" regarded it like a fallen comrade as the camera snapped their pictures. ***F–G*.**,Then they ascended and vanished into the night. (Chimineas Ranch, San Luis Obispo County, California; Sony s600)

can alert us to animal latrines, predator kills, burrows, and dens—all potential sites for camera sets. They can also save us time when searching for the haunts of unfamiliar species.[7]

Lacking canine field assistants, we must rely on our eyes to find physical proof of animal activity, mainly tracks, scat, diggings, and other more subtle signs. Nature is the classroom where you observe, take notes, and photograph animal sign, but you'll absorb it faster if you do your homework. Study and label those photos and read up on the animals that left signs of their presence.

Recognizing Habitat

Camera trappers normally set their sights on a relatively small selection of the 750 mammal species native to North America, and unlike birdwatchers, few of them keep life lists. Commonly camera-trapped mammals are widespread and abundant species, like white- and black-tailed deer, black bear, bobcat, coyote, gray and red fox, raccoons, and striped skunk. These **habitat generalists** thrive in many habitats. Less commonly photographed are **habitat specialists**. Otter and mink are familiar examples of wetland species, but habitats of other specialists are not always easy to recognize. Photographing unusual and uncommon species is a test of ecological knowledge and observational skills, especially when in terra incognita.[8]

A field guide may show you that the world's smallest rabbit—our curious burrowing pygmy rabbit—lives in big sage communities across the Great Basin and intermontane west (Figure 6.4*A*–*B*).[9] The rabbits have a patchy distribution, however, and finding colonies is another matter. Only deep but firm (often silty) soils are suitable for their burrows, which are often hidden in dense stands of sage. But their tiny "bunny beans" are unmistakable, and once you find a colony, their sign seem to be everywhere.

Figure 6.4*A*. The ever-vigilant pygmy rabbit (*Brachylagus idahoensis*). (Mono County, California, Sony s600)
Courtesy of Ken Hickman

Figure 6.4*B*. Big sage steppe, the realm of the pygmy rabbit. (Lander County, Nevada)
Courtesy of Eveline Larrucea

Many small mammals occur in microhabitats that are difficult to discern. Shrews that live in forested habitats usually confine their activities to the humus layer, rotting logs, and leaf litter, but tolerance of soil moisture differs among species. The wandering shrew (*Sorex vagrans*) and short-tailed shrews are widespread habitat generalists, while the desert shrew (*Notiosorex crawfordi*) lives in the arid southwest. Marsh and water shrews (*Sorex palustris* and *Sorex bendirii*) are the ecological equivalent of river otters on a Lilliputian scale. If you are keen to photograph different species of shrews or small rodents, study habitat in fine grain (Figure 6.5).

Habitat specialists are worthy quarry of camera trappers who enjoy challenges, but homework will save you time and effort. Field guides are a good place to start, and searchable internet sites like iNaturalist (https://www.inaturalist.org/) and Arctos (https://arctosdb.org/) give specific localities where animals have been seen or collected for scientific study. In my state, the California Roadkill Observation System is another source of information (https://www.wildlifecrossing.net/california/). Similar roadkill initiatives are underway in other states. Meeting amateur naturalists and volunteering with friendly field biologists will expose you to fountains of knowledge about wildlife. These folks know where to find elusive species and are often willing to share their knowledge.

Novice camera trappers take delight in their initial successes but will eventually want to photograph more elusive and charismatic species. Success in photographing new species depends on preparation and being in the right place at the right time. Time and patience will eventually reap new species records, but ecological knowledge can save time in getting the desired results.

ADVANCED METHODS

The inspiring work of professional nature photographers is luring camera trappers to take up artistic wildlife photography.[10] Most use DSLRs for portraits that extol wildlife in nature's realm. A smaller following seeks to document animal behavior and natural history in HD video. Both pursuits require mastery of camera technique, lighting, and timing.

A

B

Figure 6.5. Microhabitats: **A**, leaf litter and rotting logs, the microhabitat of the wandering shrew (*Sorex vagrans*) in Humboldt County, California; and ***B***, a bushy-tailed woodrat (*Neotoma cinerea*) hiding in a forgotten pile of log slabs in Flathead County, Montana. (Sony s600)

Triggering the Camera in Absentia

Animal activity will trigger photos anywhere within the camera's detection zone. If your settings are for single pictures and a long photo delay, you will get a single photo from each animal visit. If the photo delay is brief and the animal hangs around, you may get several pictures. Burst mode will also record multiple images. Generally, the more photos the better the chance of getting a pleasing portrait. If you want your camera to photograph animals in a specific area within the frame, try to restrict the size of the detection zone; use electrician's tape to reduce the size of the Fresnel lens or attach a plastic snorkel over it. This applies to commercial and homebrewed trail cameras where the PIR sensor is on the camera.

DSLR and mirrorless camera traps are often used with freestanding motion sensors having narrow detection zones. These PIR and AIR sensors can photo-capture subjects much more precisely within the scene, and with experience, you will get better at predicting behavior and movement.

Videographers want their camera trap rolling when the animal walks into the FOV. This is easy to accomplish when game trails lead to the site. When wireless motion sensors are staked on both sides of the set, the camera will start filming well before the animal reaches the FOV. Where there is no well-defined trail, use a commercial trail camera to see where animals most often arrive, and set your motion sensor along that path. Learning to live with false triggers from no-shows is part of the game. Small mammals and birds will trigger wireless motion sensors and fail to make an appearance, and large mammals may turn off the trail before reaching the set.

Using Off-Camera Flash for Portraits

The built-in flash units of commercial and homebrewed trail cameras cannot create special lighting effects. Off-camera lights, on the other hand, can imbue nocturnal settings with an evocative sense of place.[11] They highlight details and contours of an animal's body, fill shadows with subdued light, and usually prevent eye-shine. Wildlife photographers develop distinctive styles of rendering their subjects with off-camera lighting. Some strive for the dim dusky glow of twilight, while others seek the brighter effects of directional light and textured shadows (Figure 6.6*A*–*F*). In daylight, off-camera flash can brighten dark shadows, and multiple lights can emulate the ideal lighting of day when harsh light and shadows spoil the scene.

Homebrewed point-and-shoot camera traps can take stunning photos in daylight and darkness. Unlike DSLR cameras, they do not easily allow precision placement of the subject within the frame. Their slower reaction times and wide-angle PIR sensors mean that animals can trigger photos moving anywhere in the frame. But they are suitable for off-camera flash photography if they feature manual mode or adjustable flash intensity. In automatic mode, these cameras cannot meter off-camera lights and will overexpose images. Point-and-shoot cameras can trigger almost any remote flash unit in darkness if the hot shoe on the flash is fitted with an optical sensor device, previously referred to as a slave trigger (Figure 6.6). The optical sensor must be in sight of the camera, and the camera's flash must be bright enough to trigger it.[12]

Flash units modified for remote use are another lighting option. Commercial flashes can be hacked to reduce power consumption, accommodate larger batteries, and fire in response to the camera's flash.[13] The addition of dipswitches makes the remote flash compatible with point-and-shoot cameras having different pre-flash burst signals. Almost all cameras use one to three low-intensity pre-flash bursts to meter flash intensity. Such remote flash units can be left in the field for a month or longer, depending on shooting frequency.

Two or more remote light sources of differing intensity can create appealing night photos. The brighter main flash is usually placed overhead and near the camera to highlight the animal's body contours and cast shadows. If placed directly above or slightly behind the subject, the main light can highlight the animal's silhouette like a halo. The camera's built-in flash is normally weaker and fills shadows with dimmer light, while a lateral flash may be used to illuminate the background.

Testing usually has to be done at dusk when the optical trigger can respond to the camera's flash, and you can still see what you are doing. To visualize lighting effects, you'll need a surrogate subject—your teddy bear or favorite toy animal will suffice. Set the camera to force flash, and then use your camera settings to slightly underexpose the surrogate subject. If the camera's flash power is adjustable, underexpose by reducing the flash output. Next, turn on the off-camera flash units, shutter the same photo, and review the lighting effects. You probably won't get it right the first time, so adjust the distance of the remote flashes or camera settings and continue testing until your teddy bear is pleasingly illuminated with distinct but textured shadows.

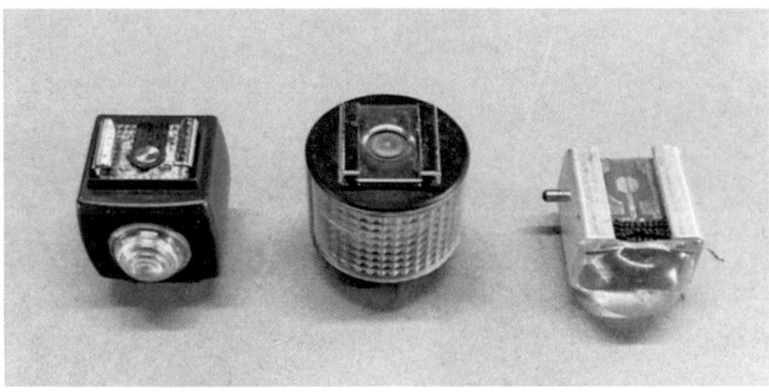

Figure 6.6. Three models of optical flash sensor. Each can be attached to the hot shoe of an external flash (or strobe) unit. When the camera's flash fires in darkness the optical sensor triggers the external flash.

Night photos taken with multiple flash units can be rendered dimly or more brightly, as if lit by twilight or lightning. Flash units may also draw attention to other features of the setting. Here on the next three pages in Figures 6.7*A–F*, some camera trappers share thoughts about their sets.

Figure 6.7*A*. "A female cougar visits a camera trap set in the Smoky River valley in Alberta, Canada. The image was made with a Nikon D3200 camera, a kit lens, a TrailMaster 1500, and three Nikon SB-28 flashes. While I sometimes like to put the subject in a more expansive scene depicting the environment, I like the challenge of matching the critter with specific elements, like this juniper bush and lone rose bud. This is where an AIR sensor, like the 1500, really shines."
—ARTHUR VEITCH, Grande Cache, Alberta, Canada

Figure 6.7*B*. "In southern California, I like to compose shots that show how close mountain lions live to a massive city. They are almost always long exposures, so the city lights can saturate the sensor. For this scene, I had been watching a few individuals travel a ridge with the broadcast towers of Mount Wilson as the background. I then realized if I turned the setup 180 degrees, I could do the same with stars and a wild scene. The camera was a Canon 6D and the lighting was two Nikon SB-28s. I used a GameWatcher PIR sensor from Ralph Davis that I narrowed to a very specific field with the use of some high-tech pieces of blue painter's tape. A bit of moonlight helped out with a natural looking cool blue cast to the scene."
—JOHANNA TURNER, Los Angeles, California

Figure 6.7*C*. "I was surprised to find a grizzly bear had stepped into a camera trap set to highlight a cool willow bush. The set was only a kilometer from a residential area. I like the color version of the image, but I chose the black-and-white version because it better highlights the texture of the branches and the bear's wet fur. A Camtraptions V3 sensor triggered the Sony A7III camera, which was fitted with a Rokinon 24mm lens. Three Nikon SB-28 strobes lit the scene."
—ARTHUR VEITCH, Grande Cache, Alberta, Canada

Figure 6.7*D*. "A mother lynx and her kitten frequented and sometimes napped on a small knoll along a ridge. The site became the focus of my set. The photo depicts the kitten playfully greeting his mother after waking her up. Dried leaves and nutlets of green alder splashed the foreground with color, and a light snowfall sprinkled the background with starlike points. The photo was snapped with a Canon EOS 5DS R and my favorite super sharp Canon 16-35mm f/4L IS lens was set at 24mm. The camera was manually set to 1/250 second, f/9, and ISO 200, and three hardwired Nikon SB-28 strobes lit the scene. TRLcam custom-built the camera housing, PIR sensor, and control board."
—KALON BAUGHAN, Helmville, Montana

Figure 6.7*E*. "The focal point here was a lakeside rock frequented and scent marked by various species. I concealed my camera trap from human visitors and placed the strobes low, which created the long shadows. The red fox was extremely wary and gave the camera a scornful parting glance as the shutter snapped the picture. The camera was a Nikon D810 camera body with a vintage ground aspherical element Nikon 20-35mm f/2.8 D AF lens set at 20mm. It was manually set to 1/200 second, f/8, and ISO 400, and I used three Nikon SB-28 strobes. The camera trap housing, PIR sensor, and control board were custom-built by TRLcam."
—KALON BAUGHAN, Helmville, Montana

Figure 6.7*F*. "Photographing bats in flight requires biological knowledge of each species and its habitat and technical skills with DSLRs and sensors. The deft use of external flashes is key to highlighting their fascinating physical features and adaptations to flight. We used three Nikon 850 cameras at a time with 50-105mm lenses set at f16. Custom-made sensors trigger the camera. To freeze the bats' movement, we set six to ten Nikon D900 flashes at 1/64th power. The sensors can trigger the camera (shutter speed 1/250 second), which in turn triggers the flashes. Or the sensors trigger a flash (other flashes on remote), and the cameras are set on open shutter (bulb)."
—SHERRI AND BROCK FENTON, London, Ontario, Canada

When equipped with wireless flash triggers, DSLR cameras offer greater freedom in flash placement than photo-optically triggered flashes (Figure 6.7). They can also trigger multiple flash units with different ratios of light. Most popular are RF triggers, which attach to hot shoes of camera and flash. The camera's commander unit (transmitter) sends signals to remote units on the flashes (receivers), and the position and distance are not limited by line of sight. High-gigahertz (GHz) models are more dependable than those with lower frequency signals. As with optical triggers, the lighting effects of multiple flashes should be tested in shade, at dusk, or at night.

Photographers use both TTL flash metering and manually programmable flash units like Nikon SB Speedlights. Generally, the camera controls its own proprietary TTL flashes, but wireless flash triggers are needed to activate Nikon and generic flash units. At present, most camera trappers use Nikon Speedlights (known for low battery consumption) in manual mode with wireless triggers.

There are always obstacles to overcome when integrating a DSLR with flash units and motion detectors. Controllers have to be programmed to periodically awaken cameras that are prone to go into **sleep mode.** Flash units and flash triggers may have to be hacked to accept larger batteries for long stints in the field (Figure 6.8). Cameras usually have to be set for multiple exposures—burst, continuous, or sports mode—because flashes that "sleep" to conserve power can miss the first photo.

Artistic wildlife photographers are known to compose photos with the deliberation of a chess player (Figure 6.9*A–C*). They use motion sensors with a narrow field of detection to photograph the subject in a predetermined place. Commercial laser or AIR sensors trigger photos when the light beam is interrupted. Their expense and power demand have roused a few gifted camera trappers to hack devices that suit their needs.[14]

A

B

Figure 6.8. ***A–B***, An off-camera flash hacked from a used Vivitar 283 flash and triggered by an optical sensor on a bendable "arm." ***C***, It is important to aim the optical sensor on the pigtail at the built-in flash of the Canon A650.

C

With camera and lens set for manual operation, wildlife photographers prefocus the imagined subject, often fix the lens's focus ring with masking tape, and arrange RF-triggered flash units for optimal lighting. They find optimal exposure by taking test shots and adjusting settings. Sometimes they use scent lure or bait to induce their subjects to pause in the right place. To avoid camera contamination, this is the final act in making the set.

Using External Lights for Night Video

Off-camera LED arrays enhance night video. To see the effect of an off-camera infrared array, position two commercial trail cameras

Figure 6.9A. Casey Smartt's diagram of a DSLR camera trap set. The camera's radio frequency transmitter triggered two remote flashes equipped with receivers. The wireless PIR sensor was aimed at the trails through a patch of prickly pear in south Texas. *Courtesy of Casey Smartt*

60 to 90 degrees to one another, aimed at the same target but at slightly different distances. Take test clips to fine-tune camera distance and angle for optimal light. Each camera's array creates shadows and improves lighting for the sister camera. Using an HD trail camera as an extra light is an extravagance, but early-model trail cameras that take inferior images can have a second life as off-camera LED arrays. The downside is that two independent cameras don't record synchronous clips, but the overlap is often extensive.

B *C*

Figure 6.9*B–C*. Casey rendered the lighting as twilight, and the collared javelinas (*Tayassu tajacu*) made their appearances as predicted. (Atascosa County, Texas) *Courtesy of Casey Smartt*

Converting DSLR and action cameras to infrared is intricate and costly, but recording night video with off-camera white LED spotlights is an alternative. Two to four 18–27-watt spotlights (designed for trucks and off-road vehicles) adequately illuminate most areas within 20 feet of the camera (Figure 6.10). They are probably less distracting to animals when mounted on poles 8–10 feet high. Most mammals habituate to them, but in my experience, owls clearly shun them. Wireless activation eliminates the need for cables and simplifies setup.

Figure 6.10. LED vehicle spotlights (18 watts) can yield complementary lighting effects for video that are similar to those created with multiple remote flash units. This one is powered by four lithium (18650) batteries.

Battery-operated home security lights can also be used as a source of white light[15] (Figure 6.11). Set a trail camera near a home security light to see the effects, and remember to set the security light's timer to run longer than the camera's video clip. The wide-angle motion sensors of security lights can detect animals and light the scene before the animal steps into the FOV.

Finally, there is another alternative. Professional wildlife filmmakers currently use high-end 4K lowlight mirrorless cameras like the Sony A7SII. With automatic ISO as high as 25,600, these cameras capture footage in daylight and moonlit nights without the need for artificial lighting. To see a variety of high-end camera traps visit Jeff Dale's blog and forum on *TRLcam.com*.

Figure 6.11. A solar-powered security light can be used as a floodlight for trail camera videos. Since the camera and floodlight are triggered by different motion sensors, the video and lighting will not be tightly synchronized.

CHAPTER 7

Camera Trap Sets in Practice

> Documenting wildlife and recording behavior has never-ending possibilities, especially in fauna- and habitat-rich landscapes. Cameras set on game trails, fallen logs, boulders, outcroppings, and even animal scat will disclose the presence of many terrestrial mammals and birds. When set at seasonal windfalls of plant and animal food, animal carcasses, or water in short supply, they reveal the importance of vital resources. For insights into social behavior, tree cavities, nest boxes, dens, and burrows offer great rewards. Subterranean and underwater sets are technically challenging but promise rare and unique perspectives. Regardless of set location, the unexpected is always possible and usually illuminating.

Camera trappers give their sets descriptive monikers. Trail sets, decoy sets, and carcass sets are handy ways of recording observable details of camera trap locations and their dominant features. These terms are arbitrary and not mutually exclusive. A scat set can also be a trail set, and in fact you could just call it a trail-scat set. Sometimes other kinds of modifiers are appropriate. A short set, for example, is one lasting only a night or two, and a DSLR set is usually an elaborate camera setup with external flash units. A ninja set is my term for a flimsy makeshift set that tempts fate. On the following pages, I discuss the challenges of different kinds of sets and share some experiences, both good and bad.

TRAIL SETS

The trail set is the camera trapper's dependable standby (Figure 7.1). Mammals traffic hiking trails, dirt roads, and firebreaks, but so do people who can be tempted to filch your camera. Animal

Figure 7.1A. Youthful exuberance. These cubs raced up the hill, but their mother took her time. (Butte County, California; video screenshot, Browning Recon Force Extreme)

Figure7.1*B*. This bobcat and kittens shared the footpath with the staff of the Audubon Canyon Ranch. (Marin County, California; Sony s600)

trails tend to track contour lines, converge at intersections, sometimes circumvent obstacles—like fallen logs and fences—often funnel animals through narrow passes, and eventually peter out. In arid lands, gullies and dry washes are heavily used as game trails, and bottlenecks in canyons have always been good places where outlaws and camera trappers have ambushed stagecoaches and wild animals. Trail traffic often ebbs and flows with seasonal changes in availability of food and cover. Most mammals are not exclusive trail users, and a few mammals seem to spend more time off-trail, plowing through dense vegetation.

Prolonged monitoring of trails can disclose rarely seen events. A well-worn trail near my house has given me video clips of playful and courting bears, striped skunks in rough-and-tumble play, pumas in estrus, gray fox and ringtails carrying half-eaten prey, and entertaining clips of a spotted skunk chasing a worried-looking gray fox.

Logs and limbs on trails are a hindrance to bipeds but rarely to wildlife and dogs (Figure 7.2*A*–*E*). I was ready to take a chain saw to an annoying log that I had to crawl under or straddle to pass. Pondering how other animals negotiated the obstacle changed my mind, and I made my first trail obstacle set. The log was a busy overpass for squirrels, an occasional perch for hermit thrushes, and an underpass for slinking deer and bears (Figure 7.2*B*–*D*). Evidently it was also scent post, because practically every passing bear broke stride to sniff it with a certain ardor. Since tall bears couldn't pass under it without rubbing their backs, they imbued the underside with bear balm even I could smell. This raised other questions about game trails, and I soon discovered that bears often jog down gentle slopes and slowly grind uphill in low gear. A young mountain lion, on the other hand, simulated overdrive and ascended a steep trail in a few leaps. I was glad I didn't take the chain saw to that log.

Trails are a good place to begin your journey in camera trapping, but eventually curiosity will lure you to other features on and off trails.

SCAT, SCENT POST, AND SCRAPE SETS

Scat sets are an easy way to photo-capture various mammals assessing the calling cards of unseen neighbors (Figure 7.3). Books on animal tracking will tell you everything you need to know about animal scat of different

A

Figure 7.2A. Trail traffic is more interesting when obstructions bar the way. (Butte County, California; video screenshot, GoPro Hero2)

species and where to find it. Your cameras will prove that scat is not a species-specific channel of communication. Multiple species pay attention and respond to the scat of other species, but not necessarily in kind. For example, foxes may dribble urine nearby after sniffing the scat of other species. Species also differ in their "scat appeal." Bear scat holds little interest for foxes, bobcats, or my dog, but cat scat is almost always a more engaging distraction.

Figure 7.2*B–D*. This series shows a black-tailed buck easing through the widest gaps of a fallen tree. (Butte County, California; Browning Recon Force Extreme)

Scat can reveal species of interest. Suppose you want to photograph otters. You can set a camera on a riverbank, or you can locate their latrines on riverbanks, lakeshores, marshes, and logjams. Otter scat is recognizable by the remains of fish, crayfish, and waterfowl. Your camera will tell you how often and what time of day they leave their calling cards, and video clips may reveal their interesting toilet etiquette (see "Scent-Marking: Leaving Calling Cards in Time and Space").

Figure 7.2*E*. This bear squeezes through a tight spot. (Butte County, California; video screenshot, Browning Recon Force Edge)

Scat and urine of male mountain lions are often associated with piles of duff called scrapes, created while squatting and treading the hind legs in a manner reminiscent of Michael Jackson's moonwalk[1] (Figure 7.4). Just like tomcats, male bobcats, lynx, and mountain lions also squirt urine on tree trunks, logs, rocks, and vegetation. Fleeting events like these can be photographed in nature. Anyone can learn to recognize trail scrapes, and your dog will point out feline spritz posts. The high-octane stench of cat pee will confirm that this is where you should set your camera.

Rubbing trees and buck scrapes are scent posts where skin secretions are the medium of communication. Bears stand upright to sniff and scratch tree trunks with claws and canines, then rub their back, upper body, or rump, which accounts for the polished patina of some habitual rubbing trees (Figure 7.5). Like other carnivores, they vigorously rub their head and neck on smelly substances wherever they find them. This leaves hair in the bark and impregnates their fur with scent. Where I live, bears commonly chew and splinter wooden utility poles, probably because they smell of wood preservatives.

Figure 7.3. Observations at a bobcat latrine in a grove of holly leaf plum at Chimineas Ranch, San Luis Obispo County, California: ***A***, a tom makes a nocturnal visit (1/7/2011, 19:34 hours); ***B***, he leaves a latrine deposit twelve days later (1/19/2011, 1:04 hours); ***C***, he appears with a freshly killed pocket gopher later that morning (1/19/2011, 9:36 hours); ***D***, he pauses to sniff the latrine with the gopher at his feet (1/19/2011, 9:36 hours). (Sony s600)

Rutting bucks scrape bare areas in the forest with their forefeet while scent-marking low-hanging twigs with the face and preorbital glands—a habit well known to hunters, who place cameras at natural scrapes and those of their own making to scope out the bucks. Scent-marking behavior of other species is more difficult to photo-capture unless you use scent lures, which are covered in chapter 8.

SCENT-MARKING: LEAVING CALLING CARDS IN TIME AND SPACE

Scent-marking is a quintessentially mammalian form of communication. General and localized skin glands give mammals their own personal bouquets that linger and inform other mammals of their presence and whereabouts. They use specialized movements to deposit these scents and to anoint themselves with botanical or foreign animal scents. They leave additional scent deposits with urine and feces. Metabolites in feces and urine contain chemically encoded personal information, which scientists can assay to identify species, determine sex, genetically fingerprint individuals, and assess hormonal profiles and reproductive status. Experimental studies show that many mammals have powers of olfactory discrimination that rival scientific instruments. Indeed, our camera trap subjects have a unique ability to read invisible events in time and space in ways we can never know.

LOG, BOULDER, AND PERCH SETS

Mammals and birds commonly visit fallen logs, stumps, boulders, and outcrops as lookouts, perches, resting places, and feeding areas. They also use fallen logs as travel routes and bridges (Figure 7.6*A–C*) over ravines and waterways.[2] Logs are usually great places to photograph rodents, which can trigger an overwhelming number of photos and fill a 32GB SD card with video clips. Scratching logs may also reward you with photos of feline pedicure behavior and other surprises (Figure 7.7).

Camera traps give unexpected insights into the minds of wild animals. I got a bright idea watching eastern gray squirrels using a split rail fence to gain access to a neighbor's bird feeder.[3] They scooted along the rails and vaulted the fence posts blocking the way. Envisioning this scenario as thrilling head-on video, I strapped a trail camera on the post and aimed it down the rail. The camera stopped the squirrels bug-eyed in their tracks. They detoured to the ground or leapt to a nearby sapling but refused to leap over the camera. And they were superstitious. When I removed the camera, they refused to leap over the fence post for several days.

A

B

Figure 7.4. Scent-marking pumas: **A**, male churns dead wood duff and soil to create a scratch pile; ***B***, female dribbles urine on a scratch pile of decayed Douglas fir created by a male. (Vancouver Island, British Columbia; Browning Recon Force Extreme)
Courtesy of Gary Schroyen

FOOD WINDFALL SETS

Nothing attracts wildlife like food, especially seasonal windfalls of nutritious food. Annual salmon migrations feed grizzly and black bears and rowdy flocks of gulls, but fruiting trees, shrubs, and vines are also wildlife magnets. The fruit of trees and shrubs is known as **mast**. Hard mast includes acorns, the nuts of hickory, pine, and beech trees, and a great variety of smaller hard-shelled fruits known mainly to small mammals, birds, and botanists. Soft mast includes berries and pomes—apples—but also buds, catkins, and even the fruiting bodies of fungi—mushrooms.

Fruiting oak trees host an impressive procession of acorn-eaters (Figure 7.8). Turkeys, band-tailed pigeons, deer, feral pigs, squirrels, and small rodents can swamp a trail camera with monotonous imagery, but the unexpected reward may be striped skunks and gray foxes deftly cracking acorns and spitting out the shells, and the commonplace clip of a strolling bear is far more interesting when it is audibly munching acorns.

Figure 7.5. Momma bear enjoys a rub-fest with an alder tree while her cubs kill time. (Vancouver Island, British Columbia; Sony s600)
Courtesy of Gary Schroyen

Figure 7.6A. The snag was an occasional night perch used by this great horned owl. (Saskatchewan, Canada; Nikon 3200 with two external flashes) *Courtesy of Daryl Dean*

Figure 7.6*B*. Gray foxes frequently use the basalt boulder near my house as a lookout. (Butte County, California; Canon A650 with two external flashes)

Lone fruit trees on old farmsteads also make rewarding sets. Bumper crops of apples, pears, figs, or plums are powerful food beacons; and since the fruit can last for weeks, camera trappers have time to learn the habits of the visitors and try different sets. Cherry trees host rowdy avian food fests that summon bird predators, like accipiter hawks and snakes, and thus are good places to photograph birds mobbing these predators (see chapter 12).

Many small mammals make a living by gathering, caching, and dispersing seeds of wild nuts and berries. Finding a rodent's food cache raises questions that resonate particularly with students of native plants. If you don't think extracting nuts from pinecones is a remarkable feat, try it yourself using only your hands and teeth. Set your trail camera at a squirrel's pinecone midden and you will witness the age-old "arms race" between cone-bearing trees and one of the world's greatest seed predators. Pinecone armaments—spines, spurs, hooks, and pitch—don't make it easy, but squirrels still manage to get a meal.

Haystacks of pikas and mountain beavers[4] and the seed larders of kangaroo rats are also busy places where a trail camera can bear witness to animals at work, as well as sneaky neighbors pilfering the cupboard (Figure 7.9). If you have curiosity and patience, your camera trap will answer many questions and pose others.

Social insects—bees, ants, wasps, and termites—also provide windfalls of animal food. The entrance to the nest is like a vending machine, a gift that keeps on giving, but a mother lode of succulent larvae and honey awaits any creature that tears into the brood chamber. Bears and skunks are quite willing to pay the price for a meal of hornets, larvae, and honey. Remains of papier-mâché hornet nests prove that it is a common event of late summer. Camera trappers don't normally set cameras around honeybees, wasps, and yellow jackets, but beekeepers and pest-control

Figure 7.6C. This outcropping was a playground for a family of Western spotted skunks (*Spilogale gracilis*). (Butte County, California; Sony s600)

Figure 7.7. A serendipitous set: ***A***, Three weeks after he pedicured his claws on the fallen log, this bobcat returned. ***B***, He found an imposter lurking in the shadows. The camera captured his shock at the moment of confrontation. (Atascosa County, Texas; Nikon D3200 with two flashes) *Courtesy of Casey Smartt*

specialists know the precautions. Nonetheless, a few videos on the internet show bears attacking hives and gray fox preying on wasps at night.[5] Hive-building in itself is a fascinating subject for time-lapse photography, though fraught with the same dangers.[6]

Ant nests are not subjects that rouse the interest of most camera trappers, but these windfalls attract specialized feeders like horned lizards, flickers, armadillos, and perhaps other species yet to be disclosed. Also known to be seen near ant nests are birds dabbing themselves with ants in a curious form of grooming known as anting.

Anyone who dismisses social insects as boring subjects should view the late Joan and Alan Root's award-winning film *Mysterious Castles of Clay*.[7] NBC was skeptical that a film about termite ecology would sell on prime-time TV, but they decided to take a chance, and two years later, the Roots' stunning footage took audiences by storm. Windfalls of natural foods offer curious naturalists many opportunities to explore fascinating feeding habits and curious ecological relationships.

Finally, flower gardens and bird feeders offer the chance to photograph hummingbirds and charismatic insects like hawk moths of the family Sphingidae. High-speed photography of insects and birds relies on sophisticated cameras and sensors, but even point-and-shoot cameras can freeze motion nicely when using flash in dim lighting (Figure 7.10). At busy nectar feeders, hummingbirds can fill a card with some pleasant surprises in just a few hours every morning.

Figure 7.8. Acorn woodpeckers adopted my neighbors' pump house for their acorn granary. ***A***, My neighbors swept the nuts out the door. ***B***, The woodpeckers promptly retrieved the acorns and ***C***, pounded them into the crack above the door, where they fell to the floor. (Butte County, California; video screenshots, GoPro Hero2)

Figure 7.9. Monitoring food caches: **A**, pika (*Ochotona princeps*) posed at its haystack on a talus slope in the eastern Sierra Nevada; ***B***, long-tailed weasel searches the site for pika. (Mono County, California, Sony DSC H70) *Courtesy of Ken Hickman;* **C**, A mountain beaver (*Aplodontia rufa*) tends to its haystack of corn lily and mountain alder clippings. (Sierra County, California; Sony s600)

CARCASS SETS

Carcasses of medium to large mammals have a powerful pull on predators, especially in winter when food is scarce (Figure 7.11). Deer that die of hunting wounds attract mammalian predators and raptors, as well as unexpected rodents, such as flying squirrels. Equally attractive are the viscera or "gut piles" of deer, elk, and moose left by hunters in the woods. Carcasses draw a succession of feeders, with the largest share going to apex predators, but scavenging foxes, ravens, crows, magpies, and jays also visit and often cache tidbits nearby. Later, cameras may record scavengers robbing these caches.

Dogs are superb at locating carrion and carcasses, but during hunting season dogs and people alike should wear blaze orange. Locating carcasses out of season is more difficult, but some camera trappers have a knack for finding mountain lion kills. If you find a kill, lashing it to a stump or tree will delay large predators from dragging it out of the camera's view. Close-up video of feeding carnivores illustrates the gory efficiency of jaws, teeth, and paws.[9] I often wondered why no cameraperson had set a camera in the thoracic cavity of an eviscerated ungulate carcass. Then I checked YouTube and found that it's already been done. Two red foxes snarfing scraps and snarling in an elk carcass makes a good show.[10] I'm sure some gonzo camera trapper will try it again.

Carcasses host a changing cast of invertebrate and vertebrate diners, depending on the state of decomposition. My friends and I set a camera at a cow elk that had died of natural causes. It looked fresh, and since the weather was cold, we expected coyotes and other mesocarnivores to fill the card with interesting pictures. Evidently it had already lost its appeal to these species. The camera recorded a lone burrowing owl and a few rodents grubbing about the corpse, which deflated with each passing day. The attraction was an exodus of fly maggots (Figure 7.12).

Smaller baits such as mice and pocket gophers trapped in garages and gardens also attract predators and scavengers, but predators will

Figure 7.10. My hummingbird feeder. By the time I finished my morning coffee, Anna's and Rufous hummingbirds had filled the memory card. They ignored the nonreflective black backdrop. (Butte County, California; Sony s600)

abscond or swallow them quickly unless you hide them and challenge the predator to work for its meal. Such tidbits attract visual predators like jays and crows, but scavengers locate food by scent. I was awestruck when a vulture found a ripe mouse carcass on a log hidden from the sky by a dense tree canopy. It was an amazing olfactory feat. Rodents killed with anti-coagulants and human foods should never be used as bait.[11]

Figure 7.11. A vigilant mountain lion stands beside a white-tailed deer kill in Lincoln County, Montana. (Video screenshot, Browning Recon Force Advantage) *Courtesy of David Neils*

My departing advice on carcass sets: roadkill would seem to be logical bait for camera trapping, but be aware of the legal issues discussed in chapter 12. Otherwise, supply your camera with fresh batteries and a large memory card because the feeding frenzy can last for days. Camera shy predators may choose to feed where light from the infrared array is out of sight. A second camera can overcome this, but you are in for a long marathon viewing the photos and clips.

DECOY SETS

People have used decoys for millennia to lure animals within killing range, and even silly-looking fakes can get a wild animal's attention. Alan Root was dumbfounded to learn that lions and rhinos treated inflatable plastic models of themselves—oversized, garishly colored, and reeking of plastic—"as the real thing."[12]

Responses to decoys differ among species. Plastic owl decoys have the power to transform songbirds into fiendish mobsters, but they also attract predators. My attempt to call Northern pygmy owls to an implausible wooden replica attracted a sharp-shinned hawk. It appeared shortly after a recording of the tooting owl commenced and then flitted around the model, but it wasn't fooled for long (see chapter 8, Figure 8.3). This was real-life evidence of a predator dynamic—offing your smaller competitors—that I knew only from books.

Decoys often generate photos of somewhat bored or perhaps befuddled hunters, such as a great horned owl perched on the back of a plastic turkey. Snapshots like that make you muse about what else transpired. I scorned myself for shooting stills when I viewed photos of a curious deer and gray fox with an old head mount of a black-tailed deer I'd hung. Video would have told the whole story.

Animal responses to their own reflections in mirrors have been important for studies of animal **cognition** (see chapter 11). YouTube has a number of entertaining videos of pets and wild animals reacting to their reflections.[13] It takes time to position the camera to frame both the animal and the mirror image, and even more to arrange white lights to video nocturnal animals without glare. Many animals don't respond to mirror images. Chipmunks may pause briefly, but what actually registers in the rodent's mind, if anything, is a moot point.

Getting a clear reaction, however, makes the work worth the trouble. My neighbors helped me make a mirror-image set at their wildlife watering station. Their own house cat found its own image terribly engaging and then looked for the impostor behind the mirror. The feral cats that came to drink were more circumspect. Some ignored or didn't see their image, and others were leery of it, but one cat eyed itself cautiously before drinking. It bolted when it raised its head and saw that the strange cat was still there. Chapter 11 discusses ethical concerns related to this topic.

Robotic decoys, a recent development in wildlife filming, may soon engage the imagination of camera trappers.[14] Until now, commercial decoys have been limited to stationary plastic models of deer, fawns, turkeys, or phony-looking furry contraptions that twist and jerk so strangely that it's amazing most animals don't bolt. Robotic decoys are remarkable likenesses of wild animals that actually move their limbs, wag their tails, walk (albeit somewhat stiffly), and make sounds.

ARBOREAL SETS

The forest canopy is a spellbinding life zone. We glimpse it through binoculars but rely on wildlife photographers and biologists for bird's-eye views of its elusive dwellers.[15] Climbing trees is not in the comfort zone of most camera trappers, but you can photograph arboreal perspectives without risking life and limb simply by attaching a trail camera to a tree trunk with a bracket. A tree-mounted camera may reward you with photos of high-speed chases by squirrels and trunk-foraging birds like creepers and nuthatches.

Use a ladder to mount your camera higher on tree trunks, and you might get head-on photos of an ascending marten, fisher, or bear (Figure 7.13). These animals don't climb trees just for the fun of it, but they do respond to natural temptations like ripe fruit, beehives, or hornets' nests. Even a garden fruit tree can give exciting results once you figure out how to mount the camera.

A

B

Figure 7.12*A–B*. Coyotes may have disdained the moldering corpses of fallen elk, but there was another item on the menu.

Figure 7.12*C*. This burrowing owl found it first—an exodus of fly maggots that was ready to pupate in the soil.

Figure 7.12*D*. Apparently a coyote licked up a few maggots, along with dried grass. (Chimineas Ranch, San Luis Obispo County, California; Sony s600)

Figure 7.13*A*. Two cameras (GoPro Hero2 and Browning Extreme) mounted on a bracket fourteen feet high in a leaning maple.

Figure 7.13*B*. The cameras were a bear magnet. (Butte County, California) Video screenshot, Browning Recon Force Extreme

Figure 7.13*C*. A ringtail in an oak tree. (Atascosa County, Texas; Nikon 3200 with two external flashes) *Courtesy of Casey Smartt*

QUEST FOR THE RED TREE VOLE

A camera trap study by the US Forest Service sparked our quest for the arboreal vole in northern California. My friend and I were soon humbled to learn that the biggest nests defied the climbing skills of Spider-Man, but eventually we found a few nests in climbable trees with low ladders of sturdy limbs. The problem was setting up the camera. We needed at least four hands to hold on to the tree and handle the tools, which called for a written protocol: climb tree, secure hip harness, raise and secure tool bag, drill hole for camera mount, etc.

Figure 7.14. Red tree vole in the wild. *Handheld photo; Courtesy of Nicholas Sobb, Northwest Ecosystem Survey Team*

It was a time-consuming but memorable venture, and the video clips were thought-provoking. One small nest seemed to be a magnet for Steller's jays, Douglas squirrels, and wood rats, but the camera failed before something destroyed the nest. Mysteriously it started working again and recorded the sad aftermath. Our hunch is that a Douglas squirrel, the cheeky predator of eggs and nestling birds, dismantled the nest while searching for a high-protein meal. More tree-climbing is in order to confirm the hypothesis, but we leave this to younger camera trappers

Arboreal mammals that rarely descend to the ground are particularly challenging subjects. My quest for the red tree vole is a case in point (see "Quest for the Red Tree Vole").[16] You can spot their large platform nests in the coastal fog-belt of the United States and Canada, but the height of their fortresses frustrated early American naturalists who dearly wanted specimens. Using climbing equipment of dubious value, they scaled gargantuan Douglas firs and caught the rodents by hand.[17]

Setting cameras in tall trees is not for amateurs. If you're not a climbing specialist, you need the help of a friendly arborist or recreational climber. Study the location with the specialist, anticipate challenges, and then develop your plan and assemble your tool kit for putting the camera in place.

Finally, nesting raptors are enticing subjects for remote cameras and are always popular on webcams. Raptor biologists use artificial platform nests to attract breeding great horned owls, eagles, and ospreys,[18] but habitual raptor nests that are also accessible can be found. CCTV cameras are particularly suitable for raptor nests because they can be remotely managed without disturbing the birds. They should be secured well before egg-laying and brooding, and twigs should be pruned before they leaf out and block the camera's view.

I have tried, so far unsuccessfully, to lure a local pair of red-shouldered hawks or ravens to adopt an artificial platform nest made of plywood and sticks. Apparently, there are better offerings in the neighborhood. The experience using a commercial camera taught me an important lesson: large trees sway in the wind much more than I thought[19] and that movement causes frequent false triggers. Raptor enthusiasts might overcome this problem by using a broken beam or mechanical trigger.

TREE CAVITY AND NEST BOX SETS

We are seldom treated to views of nesting birds and denning mammals in tree cavities or hollow logs. Here in California's Sierra Nevada, nearly two dozen species of birds nest in tree cavities, several species of bats roost in hollow trees, and three species of tree squirrels, porcupines, bushy-tailed woodrats, martens, fishers, raccoons, and black bears den in hollow trees and logs.[20]

Finding promising dens and nest cavities is not easy, but artificial nest cavities and platforms are a user-friendly alternative (Figure 7.15).[21] Common hole-nesting birds readily adopt artificial nest boxes. English bird enthusiasts were apparently the first to customize nest boxes to accommodate CCTV cameras. They now enjoy nesting behavior and the wonders of avian growth and development in their living room or study. Bird box cams are commercially available, and wireless or wired versions can be made from a wide range of CCTV cameras.[22]

Figure 7.15. Raptor nests on platforms generate endless photos and a few surprises. This osprey nest is located at the Napatree Point Conservation Area in Washington County, Rhode Island.
Handheld photo; Courtesy of The Watch Hill Conservancy

Trail cameras and homebrews are suitable for monitoring nest boxes externally. They reveal visits of nest-prospecting birds and squirrels before breeding season, capture the comings and goings of breeding birds, and sometimes reveal nest predators like snakes and jays. Fast-reacting cameras invariably capture some birds frozen in flight, and high-megapixel cameras make it possible to identify insect prey in the parents' beaks (Figure 7.16). Nesting birds however will fill a memory card with hundreds of photos daily, so expect a lot of desk time to view the catch.

Figure 7.16. ***A–B***, White-breasted Nuthatches delivered an assortment of insects and spiders to the nest. ***C–D***, They also delivered interesting in-flight photos. (Butte County, California; Canon A650)

Cruising the woods with binoculars in winter is one way to find natural cavities of hole-nesting owls, but few cavities are camera-accessible. Another way is to car-cruise back roads for calling owls. Finding nests is an acquired skill, which may even call for stumbling about in darkness and triangulating on the calls—though some owls, like the northern pygmy owl, are prone to call in daylight.[23] To find the sites in the future, it's a good idea to take photographs and record GPS waypoints and tracks.

Figure 7.17. ***A***, I found this Western Screech-Owl nest only a week before the nestlings fledged. A fiberglass pole was needed to raise the camera to the cavity. There was a lot of action. ***B***, This arrival was a balancing act. ***C***, This Jerusalem cricket was one of the few prey items recorded. ***D***, A pause in the cavity often preceded departure. ***E***, The parents took off in several directions. ***F***, Airborne shots were rarely well-framed, but they were always appreciated. (Butte County, California; Sony s600)

An owl seen in a cavity does not necessarily indicate nesting. Owls visit multiple cavities before the breeding season, and male owls may court females at several cavities before the pair settles on one. With persistence you may find a cavity that contains eggs or young. Aim the camera at the cavity six to nine feet away, study the photos to determine flight paths, and adjust camera position for optimal angle and height. In my limited experience, western screech owls were quite tolerant of a camera and even white flash, but infrared cameras are much better suited to photograph animals with large, sensitive eyes. (See chapter 11 for more on the effects of white and infrared flash.)

A few years ago, I had the good fortune to find a western screech owl nesting in a **snag**. No nearby trees were suitable to mount the camera, so I attached it to an extendable fiberglass flagpole and raised it about 14 feet to the nest hole. It swayed in the breeze like a ship's mast in a storm, but guylines dampened the movement (Figure 7.17). Framing the photo was frustrating guesswork because I couldn't see the camera's LCD from the ground. My neighbor, a former submariner, solved the problem by making a small periscope with a laser pointer. The laser's red dot simplified framing the nest hole from the ground, and the photos were certainly worth the work.[24] ("Extension Pole and Laser Periscope" in chapter 9 gives more details. Chapter 11 discusses the ethical issues of photographing wildlife and nesting birds.)

DEN AND BURROW SETS

I have a weakness for thrusting my arm into hollow logs and burrows to take pictures with a point-and-shoot camera. I used this stunt to test the gullibility of local fourth to sixth graders during our annual Trail Days event. We encounter the den tree at the end of the excursion, when attention spans are wearing thin.

"Okay, safety first! There could be a rattlesnake in there, so I'll test it with a stick," I say, as the class clamors to watch.

A

B C D E

Figure 7.18. Burrow sets for badgers: **A**, The reward of a well-planned burrow set—a badger greeting the day. **B**, We hoped the badger had settled on this burrow to give birth, but two days later she moved on. **C–E**, A punctured can of mackerel stuffed deep in the burrow summoned a display of this badger's digging skills. **A**: Lincoln County, Minnesota; Sony DSC P41; *Courtesy of Chuck Gackstetter*; **B–E**: Chimineas Ranch, San Luis Obispo County, California; Sony s600

Figure 7.19A. A black bear hibernaculum. We aimed the camera at a large burrow at the base of the rimrock and hoped for the best. ***B,*** The momma bear tried it out for size in November. ***C,*** She and her female cub pulled dry grass into the den for bedding. ***D,*** Then they waited for the long winter's sleep to begin. (Mora County, New Mexico; Sony s600)

Then I snap the photo and look at the LCD. "You won't believe it! It's a bear's den!"

Of course they moan when I show them the planted teddy bear on the LCD.

I've photographed the bowels of hundreds of potential dens in burrows and hollow logs and have yet to record a sleepy-eyed critter. But it's still fascinating to see "the dark within."

Mammals have a remarkable ability to find safe places to rest, hide from predators, and rear young. Those that give birth to undeveloped or **altricial** young den in relatively stable microclimates, usually burrows, hollow logs, caves, and recesses among boulders and under uprooted trees. Human-made structures are also used, like culverts, crawl spaces, old house foundations, derelict buildings, and farm machinery dumps. Radio telemetry studies show that many mammals have multiple dens within their home range, but they don't use them equally; weather conditions often dictate choice of den.

Marmot, prairie dog, and ground squirrel burrows are easy to find. Scope with binoculars to find rodents emulating picket posts near active burrows, and don't be surprised to photograph other species. Tunnel systems usually host a variety of vertebrate freeloaders (inquilines), like burrowing owls, smaller rodents, and skunks. Ground squirrel colonies are also a good place to get photos of weasels and badgers.

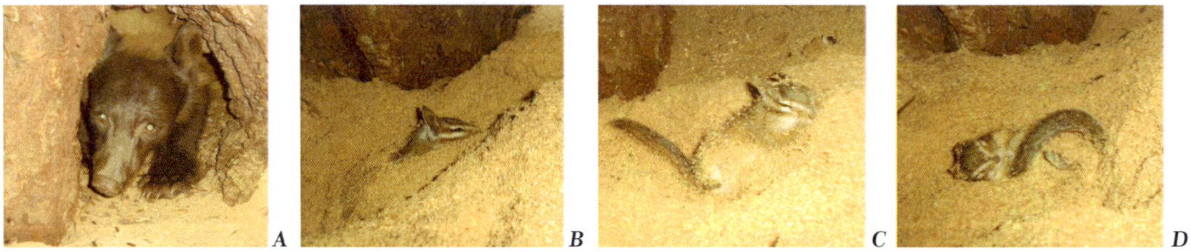

Figure 7.20. A hollow red fir infested with carpenter ants. Only a willowy student could squeeze into the opening to set the camera. **A**, The ants were at work when a yearling black bear peeked in. ***B***, The sawdust was a dust-bathing site for a long-eared chipmunk (*Tamias quadrimaculatus*). ***C***, It tossed and turned with reckless abandon. ***D***, And sometimes it napped. (Sierra County, California; Sony s600)

THE LURE OF RIMROCK AND BLUFFS

The badlands of the American West, hideaways of erstwhile "Indians" and outlaws, are evocative places. You may have to bushwhack your way in, but pleasant discoveries await you: cliff swallows nesting beneath the cornices and the whitewashed roosts of hawks and owls. Small skulls and assorted bones litter the ground, a staggering record of countless generations of nesting raptors. But the real environmental historians here are woodrats, whose presence in rocky recesses is announced by the resinous bouquet of amberat. A gang of starving forty-niners once mistook this crystallized amalgam of ancient rat urine, fecal pellets, and ratty leftovers for Native American rock candy. Now scientists study amberat to understand how plant and animal communities have changed with time.

Figure 7.21. Who goes there? A tempting rimrock cave at TNC's Dye Creek Preserve, Tehama County, California. *Courtesy of Suji Roland*

In such places you will find yourself assessing caves, recesses, and fallen rock like a real estate agent for carnivores (Figures 7.21 and 7.22). The mind spins with camera trapping possibilities, but your wildest dreams quickly vanish. You are sadly unprepared to set a camera at the raven's nest on the ledge above, and there's no way to reach that blowhole where bats stream into twilight. So you shuck your pack and settle for a horizontal cleft at the base of a cliff, where you can easily imagine a family of playful bobcats. But you are literally between a rock and a hard place. It's impossible to drive a steel stake into this bedrock, and even if you had a hand sledge, there's no headroom to raise it. So you build a cairn from the loose rock nearby and congratulate yourself for cleverness.

Making sets in places like this may demand special mounts. I enlisted my neighbor Richard to make an **expansion post** to set a camera in a rock-bottom recess that had once sheltered a mountain lion family. He produced a large supply of pipe segments of various lengths fitted with long 3/8-inch screws and nuts. You just had to link the right combination together (see chapter 9 on accessories). The invention worked, but expanding the screws was an Olympic workout. At happy hour, my frequent time-outs for wrist fatigue were a topic of amused commentary.

Figure 7.22A. A stony vault in the boulders. ***A–B***: *Courtesy of Ken Hickman*; ***A–D***: Chimineas Ranch, San Luis Obispo County, California

Figure 7.22B. Fred stood guard while I set the camera inside with the aid of a flashlight.

Locating carnivore dens is difficult in forest and brush lands; but in open range, burrows can be easily seen in the cut banks of dry washes and gullies (Figure 7.18).[25] Rimrock and bluffs are particularly rewarding places to explore (see "The Lure of Rimrock and Bluffs" and Figure 7.19). In forested habitats, incompletely uprooted trees and hollow snags are good places to look for dens—especially in winter when visibility is better and wildlife can be tracked in snow. Snap a photo inside the den with your point-and-shoot camera to get a better idea of its condition (Figure 7.20). Repeated sightings of coyotes or foxes in a small area often indicate that a den is near, and when found, it can be monitored with a trail cam. Natal dens are identifiable by fresh diggings, while tracks and scattered scraps of prey are auspicious clues of occupancy. Vocalizations of young clinch the case: you have found a natal den.

Figure 7.22C. The vault was a night roost for Townsend's big-eared bats and a rest stop for woodrats and deer mice. (Sony s600)

Figure 7.22D. Though the camera's mount somehow tilted, we were pleased with the bobcat's image. (Sony s600)

To acclimate the occupants, it's good practice to set the camera at a distance from the den. Later it can be moved closer for optimal framing. Youngsters explore the outer world and play near the den entrance, and the mother may nurse them on the burrow tailings.[26] Predators snoop around active dens and may press attacks on the occupants.[27] A denning animal's decision to move its young depends on the nature of the threat. Maternal mammals often move their litter when disturbed but may be quite tolerant in a secure den. Relocation of the litter to a different den is a piecemeal process that exposes remaining young to the threat. (More on this topic can be found in chapter 11.)

Figure 7.23. Snoozing beavers in their bank burrow. (Wayne County, Michigan) *Courtesy of Gordon Lonie*

Natal dens are often used from year to year, but game regulations may prohibit disturbances such as camera trapping during winter and breeding season. You also need permission and oversight of trained biologists to photograph endangered species at dens (see chapter 12).

SUBTERRANEAN SETS

Photographing animals in underground burrows is usually difficult, but the results are worth the trouble. Retired wildlife biologist Gordon Lonie found a beaver's bank burrow in suburban Detroit, Michigan, and noticed a small opening directly into the compartment. By placing his trail camera in the opening, he was able to photograph the family of beavers snoozing inside (Figure 7.23).[28]

Most burrows of rodents, shrews, and moles are usually too small to monitor with a trail camera, though endoscopy cameras offer a solution. Many years ago British wildlife photographer Chris Packham used this method to capture wonderful head-on footage of a European mole moving through its tunnel,[29] and more recently, video artist Sam Easterson recorded an Eastern mole boring through soil like a professional swimmer.[30]

My workshop participants helped me photograph mountain beavers in their subterranean tunnels in the Sierra Nevada.[31] Burrows measuring four to six inches in diameter are not big enough for a mountain beaver to skirt a homebrewed camera, but there is ample room in tunnels enlarged by spring run-off. We have photographed and video-recorded mountain beavers coming and going over a thousand times at several such sets. The other species that used the tunnels—long- and short-tailed weasels, mink, spotted skunk, squirrels, mice, and shrews—were a cause for celebration (Figure 7.24). Warning: mountain beavers often burrow near creeks and may have watery tunnels; during seasonal floods your camera can drown down there.

Figure 7.24. Want to photograph elusive weasels? Try looking underground in the tunnels of their prey. ***A***, Mountain beavers can be commonly shuttered within their tunnels. Their predators are also occasional users, including ***B***, mink (*Neovison vison*); ***C***, long-tailed weasel (*Mustela frenata*); ***D–E***, ermine (*Mustela erminea*) in summer and winter coats. (Sierra County, California; Sony s600)

THE CAMERA TRAPPIN' BLUES

Camera trapping is gratifying, but it can also seduce you with impossible dreams, mess with your mind, and let you down like a jilted lover (Figure 7.25). Camera trappers are known to torment themselves with wishful thinking. Approaching a set after a long wait, we wonder: Did a puma ambush an unsuspecting deer at the water hole? Did the bears get romantic on set? With giddy anticipation, they power the camera and press the view button. The LCD greets them with a chilling message: "No file in this folder." Leaving homebrews in view mode rather than photo mode is a horrifying mistake.

Mistakes happen when you don't understand the camera menu, don't check batteries, or leave your glasses in the car. Maybe you forgot an essential tool, like a rubber mallet to drive in a steel stake. So you cleverly pound it with a rock, and then discover that you just flared the metal. The sliding mount no longer fits. It's enough to make you sing the blues.

Opportunities lost brew regret, self-loathing, and longing as you trudge home in a blue funk. But the thrill isn't gone. There is always tomorrow.

A

B

C

D

Figure 7.25. Lost in the line of duty. ***A***, October 2014: The pond didn't seem to pose a threat, so I crawled into the beaver's bank burrow and found the rounded nest chamber. ***B***, September 2015: I squeezed into the tunnel, which was smaller than it was originally, and found my camera embedded in a mound of silt. ***C–D***, A skunk and a beaver paid visits two weeks after I had staked the camera. ***E***, I had thoroughly prepared to make that set, but my hope exceeded caution. Snowmelt flooded the tunnel and drowned my camera. (Plumas County, California; Sony s600)

E

Figure 7.26. The otter was a welcome addition to the packs of raccoons that roamed this Sacramento Valley marsh. (Glenn County, California; Sony s600)

WATER SETS

From ephemeral pools and seasonal creeks to springs, ponds, lakes, and marshes, wetlands offer camera trappers year-round spectacles of wildlife (Figure 7.26). Permanent wetlands and marshes host a changing cast of subjects. Marshes in spring throb with chorusing frogs and nesting songbirds, while noisy flotillas of migratory waterfowl blanket the water in fall and winter.

Aquatic mammals are usually year-round residents. Depending on the location, wetlands offer the camera trapper two species of muskrats, the introduced nutria, beavers, otters, mink, the little-known water vole, and two species of aquatic shrews. Other pleasing distractions are reptiles, amphibians, and fish, not to mention interesting invertebrates and plants that fuel the food chain.

Wetland camera trappers mount cameras on trees and saplings along banks and shorelines or make offshore sets on pipes or metal posts (Figure 7.27*A–D*). More risk-averse camera trappers anchor homemade camera rafts offshore, where muskrats and turtles climb on board to bask in the sun. At night, great horned owls may perch on the camera or dismember waterfowl in full view.

A

B

C

D

Figure 7.27. ***A***, Gary Schroyen donned waders to set a camera on a tidal creek. ***B***, He bagged a photogenic mink that posed perfectly. ***C***, His ninja set for a water ouzel tempted fate. ***D***, The set paid off before a mishap. (Vancouver Island, British Columbia; Sony s600) *Courtesy of Gary Schroyen*

Figure 7.28. I was delighted that a secretive Clapper Rail visited my seepage set in the coastal scrub. (Marin County, California ; Sony s600)

Water sets teach adventurous camera trappers that wetlands are dynamic ecosystems. The timing of spring snowmelt varies yearly and isn't found in the *Farmers' Almanac*. Mountain storms and releases of water from dams can flood distant wetlands basking in balmy summer weather. Check wetland cameras frequently, anticipate floods, and visit as soon as possible when it rains heavily.

Most birds and mammals need water daily, and isolated water holes buzz with wildlife in late summer, particularly in morning and late afternoon. Songbirds quench their thirst, bathe, and glean thirsty insects by day, while bats, owls, and nightjars make appearances after sunset. During hot spells, normally nocturnal mammals venture to drink in broad daylight. Water holes also attract predators for other reasons. At a shady spring in California's coast range, a friend and I recorded a surprising assortment of birds and took numerous photos of a furtive bobcat. It was repeatedly photographed peeking through a wall of ferns, obviously seeking a meal rather than a drink (Figure 7.28).

Shady creeks and pools are heat sinks where overheated bears cool off and groom.[32] They arrive in a cloud of gnats and flies for a sitz bath, often luxuriate in the pool while snacking on thirsty yellowjackets, and even doze off with periodic fits of scratching. When male elk approach rutting condition, water holes become social centers where little etiquette prevails beyond the rules of the dominance hierarchy. Stags strut like kingpins, display antlers, self-urinate, spar, thrash vegetation, and wallow, converting a pristine pool of snowmelt into a reeking mud hole (Figure 7.29).[33] Rutting moose dig personal urinals where wallowing enhances their natural bouquet.[34] Often cow moose visit the sites on the sly, presumably to size up prospective mates.

NINJA SETS

Sometimes a technical difficulty, like a special mount or a missing tool, prevents proper completion of a set. You could postpone the set, but that nagging little voice says, "You can't wait. This is the chance of a lifetime. A wolverine might arrive with the first dusting of snow." With a pruning saw and a Swiss Army knife, you fashion a stick or a sapling into a post and lash the camera to a cross piece with string, wire, duct tape, or bungee cord. I call the jury-rigged solution to this dilemma a *ninja set*. It's a risky exercise in blind faith and making do with whatever you find in your pack. Sometimes you return and find your camera intact, but in my flights of fancy the wolverine has yet to make a showing.

Water troughs in cattle country may lack the photographic appeal of springs and natural water holes, but they are still excellent places to survey birds and wild mammals. Be aware, however, that some cattle fixate on trail cameras as rubbing posts or lick them like lollypops. Elk are similarly inclined, and those in hard rack often regard a camera on a post as an invitation to a tilting match. Guess who wins.

Figure 7.29A–*B*. Wallows have much to offer large mammals and camera trappers alike. A bull elk luxuriates in the spa. (Vancouver Island, British Columbia; Sony s600)
Courtesy of Gary Schroyen

UNDERWATER SETS

One spring morning I surprised a family of otters sunning on a creek bank. They quickly dropped into the current, swam downstream, and vanished in a clump of Indian rhubarb on the far bank. It was a thrilling encounter, and discovering the submerged burrow was a bonus. The memory of their shadowlike forms gliding underwater periodically nags me. What a thrill it would be to photo-capture the aquatic sports of otters, muskrats, and beavers. Would water shrews tackle a hookless fishing lure or a rubber minnow? These may not be topics that keep camera trappers awake at night, but they are worthy challenges that test their mettle.

Almost everything that swims, from aquatic insects to migrating salmon, is fascinating to see underwater. Fishermen use viewing systems called fish cams to satisfy their curiosity about lunkers in the depths. There are a number of waterproof cameras on the market, but there's a dearth of underwater videos of aquatic mammals on YouTube and Vimeo. The best clips seem to have been filmed in person, presumably because deploying an underwater camera trap is more challenging than snorkeling with a handheld camera. A remote underwater camera could capture a great deal of footage with less effort and no danger of being bitten by territorial otters, but there's also a greater risk of mechanical failure.

Figure 7.29C. A feral hog wears lipstick. (San Luis Obispo County, California; Sony s600)

Figure 7.30. This two-camera system was designed to record animal activity in a Nebraska marsh both above and below the surface simultaneously. The PIR sensor and controller are at the top of the pole. *Courtesy of Jeff Dale*

Waterproof housings and underwater sensors are necessities for underwater cameras. PIR and x-band sensors don't work underwater, but CCTV cameras using pixel change detection have possibilities. False triggers can be expected from floating debris in rivers but would be less likely in lakes. Jeff Dale of TRLcam, however, designed an ingenious solution to photograph beavers and otters underwater (Figure 7.30). Knowing that a PIR sensor can only detect movement above water, he rigged two GoPro cameras powered from a battery box and solar panel on a pole. When those beautiful swimmers appeared on the lodge, the sensor activated both cameras and simultaneously filmed them slipping into the water and swimming away.[35]

Fieldwork with camera traps always raises questions about nature and rewards patience with insights and knowledge. I have discussed a variety of camera trap sets, but I like to believe that curious naturalists will experiment with new kinds of sets and find new windows into nature's secrets. The work of gifted wildlife photographers and TV productions are ongoing sources of inspiration. Critter cams and robotic decoys have given us stunning perspectives about animal behavior, but it's reassuring to know that even recreational camera trappers have brought remarkable images and footage to the public's attention.

CHAPTER 8

Wildlife Attractants

Scent lures, baits, animal decoys, and sound recordings attract wildlife to camera trap sets from long distances and thus increase the probability of photo-capture. They also elicit interesting behavior and prolong an animal's dwell time in front of the camera, often increasing the number of exposures. Attractants differ in directionality. Sound propagates from a source in all directions, but scent dispersal is usually directional, influenced by weather, air movement, and topography. Visual, sound, and scent lures are useful aids that give insights into the cognitive world of animals.

I have mixed feelings about the endearing way my dog surrenders his fetching stick. He usually drops it next to a pile of "dog logs." It's like a bad joke, but it's a puzzling habit. Does Fred drop his stick because the dog piles are distracting? Does he use dog piles as olfactory waypoints? Or does he want to share his discovery with me? Insight into the cognitive world of wildlife was a survival skill of our hunter-gatherer ancestors, who knew when, where, and how to find and catch wild animals. But they also knew how to deceive their quarry using bait, decoys, and callers. This ancient craft hasn't changed. Camera trappers use bait, decoys, and lures to attract animals and prolong their visits to sets, while field biologists use scent lures and baits as an aid to survey carnivore populations. In general, attractants increase photo-captures.

There is overlap in the use of these words, so let me define them here.[1] **Baits** are live, dead, or processed foods that are used to attract and catch wildlife. **Decoys** are fake animals (or plants) that attract wildlife. By this definition, the angler's fishing lure is a decoy. I reserve the term **lure** for attractive scents, sights, or sounds that are not foods. Camera trap lures include animal scat, urine, scent-gland secretions, feathers, artificial visual attractants like DVDs and wildlife recordings.

Mammals typically react to scent lures by sniffing, but lures can also elicit tasting; scent-marking with urine, feces, or scent glands; and rolling and rubbing (Figures 8.1 and 8.2).

Figure 8.1. Gray foxes demonstrate their three main scent marking activities: ***A***, urination (on a mountain lion scat); ***B***, defecation (near a dab of castoreum and cat mint); and ***C***, scent-rubbing the neck and cheek (on castoreum).
A: Napa County, California; ***B–C***: Butte County, California; Sony s600

SOME BOTANICAL SCENT-LURE OILS

These plant extracts can be used to attract various species of mammals.

- Almond
- Apple extract
- Anise
- Asafetida
- Bergamot
- Birch
- Catnip
- Cherry
- Cumin
- Fennel
- Pennyroyal
- Poplar bud
- Spearmint
- Wintergreen
- Valerian root

Commercial scent lures are made from a variety of compounds, including botanical extracts, scent glands and secretions of trapped furbearers, or chemically synthesized animal musks (see "Some Botanical Scent-Lure Oils"). Other products are mixtures of foods and scent glands, which blur the boundary between baits and scent lures.

Some recreational camera trappers use attractants selectively, depending on circumstances, while others use them routinely. Still others prefer to go it without. Your own philosophy of camera trapping may dictate whether to use them or not, but attractants are helpful aids when used with discretion. Biologists and citizen scientists in particular often use attractants for wildlife surveys. Regardless of your purpose, if you use attractants, you should be aware of the ethical and legal ramifications, which are discussed in chapters 11 and 12.

VISUAL LURES

Two guys in a car peer at a map with a flashlight. The shortcut didn't pan out, and now they are lost on a country road in the middle of nowhere. As they bicker about their predicament, a snarling mountain lion leaps onto the hood and paws menacingly at the nerds inside the windshield. Miraculously, the driver distracts the cat with his laser pointer, and it gamely chases the red dot into the woods. "I can't believe that worked," exclaims the driver with relief.[2]

This wacky television commercial celebrates mindless feline vulnerability to laser pointers. Cats—and visual predators in general—have an uncanny ability to detect prey movements and visual cues (see "Search Images: Aids to Seeing the Unseeable"). Bird feathers and reflective flagging rippling in the breeze can tickle their curiosity and trigger their stealth mode of exploration. Even CDs dangled from a tree are said to invite bobcats to explore.

Figure 8.2A. A black-tailed doe urinates on its tarsal glands—one of several scent glands used by this species in communication.
Courtesy of Gary Schroyen

Another visual lure is the fur trapper's classic dirt hole set, which is simply an imitation of the diggings made by coyotes, foxes, badgers, and skunks. Dig a hole with a narrow trowel and scatter the soil as a tailing. Your creation will look like the real thing. Stuff some scented wool down the hole, and predators will have an added olfactory incentive to investigate.

Figure 8.2*B*. A buck lip-curls in response to an unknown chemical stimulus. (Vancouver Island, British Columbia; Sony s600)
Courtesy of Gary Schroyen

Realistic animal models or funky garden decorations of deer and fawns also attract predators. Motion decoys, a recent development for predator shooters, are furry motorized models with little resemblance to a real animal. Visual decoys don't fool most predators for long, but they may give you an interesting photo or two (Figure 8.3). Nowadays wildlife filmmakers use realistic robotic decoys with cameras to elicit fascinating animal behavior.[3]

Decoys of owls, hawks, and snakes commonly elicit the mobbing behavior of songbirds, a fascinating spectacle usually seen during the breeding season. Alarm calls rally other birds to join the fracas, and the mobsters harass the predator with daring aerial maneuvers. While live predators seem to energize the mobsters every time they move, birds soon lose interest in motionless decoys.

Birders know that fence posts, snags, power poles, and certain tree limbs are favored perches for birds, and good places to snap photos. Some perches are also suitable places to set trail cameras. Artificial perches are also good bird attractants, especially in areas where natural perches are in short supply.[4] They can be fashioned

Figure 8.3. This Northern Pygmy-Owl decoy and recording failed to fool a sharp-shinned hawk. It took its leave after the camera made two exposures. (Butte County, California. Sony s600 IR conversion)

Figure 8.4. A bobcat eyes a digital chirpy bird caller hidden in a hollow log. (Butte County, California; Sony s600)

on the spot or prepared ahead of time for easy deployment.

Many mammals find the trail camera's "face," electronic flash, and infrared array worthy of scrutiny. Light-reflecting objects underwater can engage raccoons and otters in lively investigation. It seems that no one has used a small blinking light as a photographic lure, but there's good reason to believe it might work.

A mirror can give amusing insights into animal cognition. Charles Darwin discovered that Jenny, the London Zoo's orangutan, seemed "astonished beyond measure" when she gazed at her own image in a mirror.[5] Since then, psychologists have found that mammals and birds react to their own mirror images in different ways. Many of us have seen birds attack their images in windowpanes and even the side-view mirrors of parked cars. Some become so obsessive that covering the mirror seems the only way to give them peace of mind. (See chapter 11 for a bit more about mirror image recognition in animals.)

SEARCH IMAGES: AIDS TO SEEING THE UNSEEABLE

Mammals and birds easily find hidden and camouflaged trail cameras, but they also find prey that blend perfectly with their backgrounds. When cryptic prey like moths, caterpillars, and grasshoppers are abundant, birds encounter them accidentally or by methodical searching and quickly learn to recognize key visual patterns. The **search image** may only be an abstract marking, like a dark wing bar on a camouflaged grasshopper, but it is all they need to know to detect a meal. Experiments have shown that birds have difficulty cracking the "camouflage code" of uncommon cryptic species because they encounter them infrequently and presumably don't learn or remember the key features. Predators in general zero in on sights, sounds, and smells associated with their prey. The search images of mammals probably include visual, scent, and sound cues that give away the prey's presence, while tufts of fur, scattered feathers, diggings, burrows, scat, and airborne scent from tracks and fur are all incidental cues.

SOUND LURES

Unlike scent, sound doesn't need a breeze to spread. It propagates in all directions. Digital predator callers sold by sporting goods stores are the sound lures of the hunters' choice, and the most commonly used sounds are distress calls of prey animals such as mice, rabbits, and birds. Recordings of bleating deer fawns and elk calves are also meal calls to hungry predators. Hunters bugling for bull elk have occasionally summoned a hungry mountain lion or grizzly bear. A simple recording of a chirping bird can induce a predator to seek the source and could result in a unique photo (Figure 8.4).

Figure 8.5. Decoys with recordings: **A**, This great horned owl took measure of the dubious hooter. *B*, Though the camera failed to record the deed, the owl evidently laid low the brash intruder. The set was made in late spring after the nestlings had fledged; a hidden iPod hooted at intervals under the decoy. The owl's red-eye was color-corrected. (San Luis Obispo County, California; Sony s600)
Courtesy of Ken Hickman

Many other sounds, even squeaky toys and ticking clocks, may attract mammals and birds. Some predators may wait patiently for the mysterious caller to betray its presence, but animals usually don't hang around when they discover that reality doesn't match expectation. They seem to know that poor or unnaturally loud recordings are bogus. It's always a good idea to chain a sound player to a stout limb, as predators may abscond with it despite its obvious inedibility.

Long-distance calls of wolves and coyotes hasten pack reunions and rally group members to join the chorus.[6] Recorded calls may prompt a vocal response or draw animals closer to the caller and camera[7] (Figure 8.5). Many female mammals call persistently when in heat. Mountain lions utter a plaintive rasping call. The sound isn't loud, but it's repetitive and can "electrify" the neck hair of tented campers or even cause an accident in the sleeping bag. One evening last summer, a caterwauling lion roused my dog, who was asleep next to the screen door. The resulting commotion soon had us bellowing, "Quiet." The "love song" faded in the darkness and presumably set off the neighbors' dogs later that evening.

Sound lures might prove an effective attractant to charismatic microvertebrates like shrews. These ravenous predators seek out small vertebrates and insects by their sounds, scents, and vibrations.[8] The curious camera trapper who investigates this subject will have fun throwing light on one of nature's mysteries.

Birdsong has a territorial function. It compels neighboring birds to declare their presence with song. Ornithologists learned many years ago that recordings of hard-to-see species enticed the birds to show themselves. Song playbacks work best in the breeding season, but territorial defense is energetically expensive, especially when food is scarce during cold snaps in early spring. This is why playback of territorial birdsong should not be used as a camera trapping aid during the breeding season.[9]

Field biologists play territorial calls to survey owl populations.[10] But during the breeding season, owl calling can also disrupt foraging and the care of young, so birders and ornithologists rightly consider it as harassment. Camera trappers should use this method with discretion, preferably with advice from an owl biologist. (Chapters 11 and 12 discuss legal and ethical concerns of using sound recordings.)

A

B

C

Figure 8.6. ***A***, A bobcat urine sprays a rock. (Kern County California, Pentax Optio E50) ***B***, A bobcat urine sprays a rock that covers a punctured can of fermenting mackerel. The urine mark was evident when the cat strolled away, and the speckled lens suggests the cat may have sprayed the camera first. (Chimineas Ranch, San Luis Obispo County, California, Sony s600). ***C***, Bobcats often target shrubby vegetation to spritz urine. (San Mateo County, California; Pentax Optio E50) *Courtesy of Ken Hickman*

For thc do it yoursclfcr, construction plans for callers can be found on hunting and camera trapping forums.[11] Inexpensive electronic record–play modules can be modified as wildlife callers, and the internet is a ready source of animal sound files. An iPod or CD player also can be wired to a battery-powered speaker. When sounds are played at ten-to-twenty-minute intervals, the power will last for several hours.

SCENT LURES

Olfaction, the sense of smell, is central in the lives of nearly all mammals. By scent they find food, avoid predators, defend themselves, and communicate information about their sexual identity, reproductive condition, age, diet, and health. Urine, feces, saliva, and scent glands contain social and reproductive information. Defensive scents, like the familiar stench of skunks, have an antipredator function, while some scents serve as alarm signals to other members of the same species.

Animal trappers concoct commercial scent lures from scent glands and their musks. They commonly add animal urine, additional scents, and aromatic plant substances to "grab attention" and enhance "calling power."[14] Animal lures and baits smell revolting to most people, but they have a magnetic and sometimes enchanting effect on other mammals. Sniffing may precede urination, defecation, and other forms of scent-marking, or a rapturous bout of rolling and rubbing (see "Why Do They Do That?"). Mammals often taste scent lures in small amounts, but as a rule they don't devour them unless they contain meat and whole tissue.

Essential oils from various animal musks have traditionally been important ingredients of perfumes (see "Musks and Essential Oils").[15] Now they are synthetically produced, and demand for the natural products has diminished. Perfumes, deodorants, men's cologne, and aftershave usually contain synthetic musks that evoke interesting behavior in mammals.

The distinction between commercial scent lures and baits is not always clear. The recipes are regarded as trade secrets, and ingredients are rarely listed in detail. Sledge Hammer, Gusto, Moneymaker, Dinner Bell, Coon Pie, and Tail Slapper are just a few of their fanciful labels. Commercial trapping baits are also sometimes a mixture of scent lures, fermented animal tissues, and botanical food extracts. Some products called lures are mixtures of scent-gland secretions and

THE MAGICAL POWERS OF CASTOREUM

Castoreum, the gummy contents of the beaver's castor sacs, is probably the most celebrated mammalian musk. Its place in history goes back at least to the fourth century BC, when Hippocrates alluded to its curative powers. By 1685 a European treatise attributed medicinal properties to many other parts of the beaver's body. When beaver trapper Osborne Russell suffered arrow wounds in the hip and thigh from Blackfeet Indians in 1838, he treated the wounds with salt water and a salve of beaver's oil and castoreum. It eased the pain and drew "out the swelling in great measure," he wrote. The anti-inflammatory qualities of castoreum come from salicylic acid, which beavers get from eating willow leaves and bark.

Castoreum is a rich mixture of at least forty-five aromatic compounds derived from plant foods and is an irresistible attractant to other beavers. Biologists postulate that making mud mounds and scenting them with castoreum and urine is the beaver's nonviolent way of announcing its occupancy to invading **conspecifics**.

Fur trader and explorer David Thompson believed that early European colonists learned about castoreum's power of attraction from Indians of the northeast. Trappers soon discovered its potent attraction to other furbearers and stored the secretion in small wooden boxes or vials worn around the neck. They dabbed small amounts of the lure onto a sprig of willow or a handful of mud thrown on the bank beside a submerged steel trap.

Fragrantica, an online encyclopedia of perfumes, describes the aroma of castoreum as animalistic and sharp, "spreading a tar-like note that reminds of an odor of birch tar or Russian leather." Some Sioux Indians evidently perfumed their hair with it. Not long after the Crow Indian scout Plenty Coups scalped a Sioux warrior, his partner Big Shoulder offered the scalp for a sniff. It was scented with castoreum, which Fragrantica says is "wild and bodily, lustful and passionate, bestowing the one who wears it a delicate aura of sensuality."

When British chef Jamie Oliver announced on the *Late Show with David Letterman* in 2011 that castoreum is used "in cheap strawberry syrups and vanilla ice cream," soccer moms rushed to their computers for reassurance that it wasn't so. A ribald reference to food additives—"Beaver Butts Emit Goo Used for Vanilla Flavoring"—in American news media noted that the Food and Drug Administration (FDA) classifies castoreum as GRAS: generally recognized as safe. Snopes.com refuted some of the scary claims, but in 2013 the Swedish National Food Agency reported that castoreum "can be used to provide a taste similar to vanilla in baked goods and sweets."[12] Swedes aren't as squeamish as Yankees. They've been enjoying a beaver-butt-flavored vodka called *Bäverhojt* for some time.[13] *Skol!*

fermented flesh, fish, or shellfish oil. So-called scent-baits, call-foods, or food calls are designated as long-distance attractants.

Feces and urine are the last category of olfactory attractants. Bobcats and mountain lions habitually scent-mark bushes, tree trunks, and rocks with urine, and commercial cat urine can be used to stimulate this behavior (Figure 8.6). The latrines of many mammals are good places to make sets, but those of some species can be hard to find. Raccoons and ringtails, for example, often make latrines on elevated stumps, or on top of nest boxes intended for wood ducks. Camera trappers sometimes use scent lures to refresh the olfactory appeal of old scat. I once used my own dog's anal scent for this purpose, and guess what? A coyote stopped to sniff the scat. Did my doctoring of the scat make a difference? Maybe, but only a controlled experiment could prove it.

Baits, scent lures, and predator urine and feces can be purchased online from trapper supply stores. If you don't want to use lures from trapped furbearers, fermented fish fertilizer sold in garden shops works well as an alternative. Veterinarians often treat dogs with infected anal glands. Samples of these potent scents often can be obtained for the asking, though you risk being labeled as a wacko.

GENERAL RESPONSES TO SCENTS

A camera set at mountain lion scat soon records various species of predators before the big cat makes an appearance. Predators sniff the scent of prey species, and herbivores are attuned to the odors of predators. Their worlds are rich scentscapes that inform them about foods, predators, and members of their own species. Time spent sniffing any particular scent differs, but long sniffing times probably indicate biological relevance and chemical complexity. It may also be necessary for the sniffer to fix an unfamiliar scent in its olfactory memory.

Food habits strongly influence scent preference. Plant-eating mammals are more responsive than carnivores to botanical odors, and carnivores seem more responsive than herbivores to animal-based odors. Sniffing often precedes other behaviors. Foods of course lead to salivation and feeding, but many animals lick aromatic substances, sense it with the **Jacobson's organ**,[16] and finally rub and roll in it. Though odors of predators like mountain lions and wolves do not cause prey species to take flight, these odors may increase alertness and disrupt certain behaviors.[17] A small number of substances, most notably ammonia and capsaicin, have an immediate repellent effect on many mammals. The smell of millipedes and stink bugs, which produce a variety of antipredator chemicals, have a similar effect on my dog.

Species differ in their affinity to various scent lures. When targeting a specific species, it's a good idea to know its scent preferences. An investigation of Canada lynx found that a mixture of castoreum and catnip was a more effective attractant than four brands of commercial lures designed for wild cats. The researchers found that bears were most responsive to one of the commercial lures, and coyotes responded to all, with no apparent preference.[18]

In a South African study of brown hyenas, researchers identified preferred attractants before undertaking field surveys. They conducted cafeteria trials with hyenas in a zoo, testing thirteen substances, including canned cat food and Calvin Klein Obsession aftershave. The hyenas' reactions showed that four attractants were of greatest interest. The researchers took this as sound advice for their field survey.[19]

MUSKS AND ESSENTIAL OILS

Musks are aromatic glandular secretions that function in communication, reproduction, and predator defense in many species of mammals. The word *musk* comes from the Sanskrit word for testicle, apparently a mistaken reference to the scent gland of the musk deer of the Himalayas. Musk glands are usually skin glands, and many also have storage sacs. Raw musk is a highly aromatic mixture of compounds known as essential oils. They have great staying power, but the importance of essential musk oils to the perfume industry is the unique chemical property of "exalting" or enhancing the smell of other aromatic compounds. Skunk scent, castoreum (beaver), muscone (muskrats), and civetone (civets) are well-known musks.

Odorous plants contain aromatic oils in wood, bark, buds, and leaves. The oil is the carrier of the "essence" of the plant, hence the name *essential oils*. Steam distillation is the method most frequently used to separate the volatile oil from the plant but steeping in alcohol also works. Turpentine and oils of cedarwood, eucalyptus, and citrus trees are well-known products, but there are many others.

It's hard in English to describe smells except in very simple terms, so perfume makers invented various classifications. Musks fall into the category of animalistic odors, while the term *fragrance notes* relates to the timing of scent dissipation. Top, middle, and low notes refer to scent components that are early, delayed, or slow in the dissipation of molecules.

SPECIFIC REACTIONS TO SCENTS

Mammals, and some birds, often follow scent gradients to their sources. The posture of dogs testing the air is unmistakable.[20] Many of us have seen our dogs take fright, raise their hackles, or bark upon sniffing an unseen foe. Mammals also sample less-volatile compounds by licking and sensing with the Jacobson's organ (Figures 8.7*B* and 8.7*C*).

These facts have implications for wildlife photography. Some substances bring on a bout of rolling and rubbing, which anoints the animal's fur with the foreign smell and transfers its own scent to the site (Figure 8.7). Catnip (*Nepeta cataria*) is a familiar example.[21] The active compound, nepetalactone, is produced by several species of mint. It's a powerful cat attractant that rouses a transcendental interlude of biting, drooling, rubbing, rolling, and sensual play. Castoreum, valerian, and other oils evoke similar behavior in other carnivores, including bears and foxes (see "The Magical Powers of Castoreum"). One of the more bizarre forms of scent anointing is the Virginia opossum's saliva bath.[22] The animal licks a smelly source of scent, slobbers, rubs its neck on the wet spot, then licks its flanks until they are thoroughly saturated.

A

C

B

D

Figure 8.7. ***A***, Cats in general react to catnip and castoreum by sniffing; ***B***, licking and drooling; ***C***, lip-curling (the Flehmen response); and ***D***, rubbing the upper body on the scent. ***E***, This puma's performance ended with urination.
A, ***B***, and ***E***: Butte County, California; Sony s600; ***C***: Plumas County, California; video screenshot, Browning Recon Force Extreme; ***D***: Kern County, California; Sony s600; *Courtesy of Ken Hickman*

E

BAITS

From whole carcasses to the artificial baits concocted by trappers, baits are effective animal attractants (Figure 8.8). Carcass appeal declines with decomposition, but commercial baits with preservatives last much longer.

Figure 8.8. A black-tailed buck examines a partially consumed vehicle casualty off a country road. Is this morbid curiosity or recognition of a companion? (Butte County, California; Sony s600)

Most rodents and rabbits feed on a variety of plant tissues, from seeds and pinecones to buds and flowers. They also respond to plant extracts. It may seem odd, then, that rabbits and rodents at times feed from animal baits; but all animals need protein, and when normal foods don't supply it, they take it where they find it.

Trappers and some camera trappers use commercial baits and bait-scent cocktails. Some are said to appeal to a spectrum of species, while others are deemed to have more specific calling power. Bait ingredients include fish and shellfish oils; fermented meats and organs; or concoctions of foods, scent glands, and glandular secretions. Skunk musk is a common seasoning used in trace amounts. When the brewing mixture acquires a suitable taint, the *schtinkmeister* adds sodium benzoate to preserve the bouquet. Field biologists have used raw meat or canned fish as bait for camera trap surveys of carnivores. A punctured can of mackerel or tuna releases a succession of increasingly potent scents with amazing staying power. Wired to a tree, it will outlast most carcasses because it can't be eaten. Bears may dislodge it, but they soon recognize the futility of their efforts. Pet food should never be used as wild animal bait (see "Guidelines on Using Bait").

WHY DO THEY DO THAT?

Why would an animal anoint its body with a strong or even vile-smelling foreign scent (Figure 8.9)? Does a strange or fetid scent make the animal more appealing to its companions, as perfume does in humans? Or does it mask the animal's personal odor?

Wildlife biologists recently proposed that donning the scent of your enemy could be protective. With the use of trail cameras, they noted that gray foxes commonly rub their face and forebody in scent scrapes made by mountain lions. It is well established that larger predators like coyotes kill smaller competitors like red and gray foxes when they can.

Is scent anointing a deceptive antipredator ploy? If a fox smells like a mountain lion, can it throw a mountain lion off its trail? Could a dose of lion scent protect a gray fox from competition with and predation by coyotes? It's an intriguing idea, but only experiments can prove or disprove the hypothesis.

Figure 8.9. A gray wolf scent-rolls in an unknown scent. (Vancouver Island, British Columbia; Sony s600) *Courtesy of Gary Schroyen*

SCENT DISPERSAL FACTS

While humidity, air movement, and topography strongly affect the dispersal of scent, we have learned a great deal about the mammalian sense of olfaction from sniffer dogs employed in drug and cadaver detection and search and rescue.[23] Like most mammals, dogs can smell volatile molecules and detect trace amounts of scent in dead skin cells and skin secretions such as sweat and sebum. They detect scent more easily on still days when the weather is cool and moist. In such conditions, scent molecules disperse in all directions from the source. In a breeze, scent moves downwind in a cone-shaped zone, and its intensity diminishes with distance from the scent source. This is called the **active space.**[24]

Scent disperses farther when the source is off the ground, and this may explain why many mammals leave elevated scent marks. Raccoons, ringtails, and gray foxes often leave scat on rocks and logs; cats spray urine on vertical surfaces of tree trunks, bushes, and rocks; and bears commonly rub and chew tree trunks in an upright posture. Camera trappers can aid scent dispersal by putting scent lures several feet above ground. Another dab of lure on the ground will hold the animal's attention while the camera takes photos. Chapter 9 (Figure 9.21) shows a simple aerial scent-delivery device made from PVC.

Topography strongly influences air flow (**convection**) and scent dispersal. When rocks and logs obstruct airflow, eddies cause scent to accumulate on the far side of the obstruction. Large obstructions can divert the flow vertically for a considerable distance, above the animals' range of detection. Time of day also affects the direction of air movement, especially on slopes. Prevailing breezes move downhill at night and uphill during the day. The interaction of wind and topography can create eddies and updrafts that can throw even a seasoned scent dog off course. Understanding the fundamentals of scent dispersal gives insight into mammalian behavior. Mammals are often skittish in windy weather because the scentscape is unpredictable.

GUIDELINES ON USING BAIT

- Know your state's game code and whether the use of baits is legal.
- If baits can be used legally, limit the amount used so that the normal foraging behavior of wildlife is not disrupted. Creating large amounts of bait may be considered illegal by state wildlife authorities. Do not create food dependency in wild mammals.
- Animal carcasses are legitimate attractants at the location you find them, but you cannot transport carcasses of game animals without a hunting license and a tag, and in some cases, you may need a scientific collecting permit.
- Do not use human food as bait; it is bad practice and often illegal.
- Place baits away from areas that will attract dogs or people or place wildlife in harm's way, such as near houses, highly trafficked roads, and highways.
- If you wish to use baited camera trap stations to survey wildlife, apply for a scientific collecting permit from your state game department, and, if applicable, get the landowner's permission.

CHAPTER 9

Camera Mounts and Accessories

Buckle straps, cable locks, and screw mounts are standard means of mounting commercial trail cameras in wooded habitats. In treeless landscapes, metal or wooden posts, skid mounts, or cairns are alternatives. Steel security cases are essential accessories where bears and people pose threats. Users of homebrewed point-and-shoot and DSLR cameras rely on tripods and ball-and-socket mounts, as well as homemade mounts and security boxes. Unusual settings and perspectives often call for special mounts, like telescoping poles, expansion posts, and camera cranes. Sooner or later, you may encounter promising circumstances but lack the ideal mount. That's when on-the-spot woodcraft with an axe, a knife, and some cord may be the best solution.

The camera mount, which secures the camera to a substrate, is an essential trail camera component. The ideal mount is simple to use, versatile, and secure. Commercial trail camera mounts (buckle straps and screw mounts) meet these criteria in wooded habitats, but posts are needed in deserts, prairies, alpine meadows, tundra, and chaparral. Homebrew users with off-camera flash units and sensors usually need a variety of mounts and hardware, and those inclined to make unusual sets have to buy or build special mounts.

There is no truly universal mount, and choice of mount is often a compromise between the need for security and the freedom to place the camera in the ideal situation. Let us first consider the usual kit of accessories used for commercial and homebrewed cameras, and then describe some homemade contraptions that have solved special problems. They can certainly be simplified and improved. The forums will also give you many ideas.

Figure 9.1*A*. A nylon strap is a convenient fastener, but it will not deter a thief.

Figure 9.1*B*. When locked to a tree with a cable, a metal security box is relatively safe. *Courtesy of Ken Hickman*

Figure 9.1*C*. A DIY screw mount can be made from two eyebolts, a nut and bolt, and two washers; use a screw eye to attach it to wooden posts or trees.

COMMERCIAL TRAIL CAMERA MOUNTS

Commercial trail cameras are usually designed for three ways of mounting. First, cameras and their security boxes can be belted to tree trunks using the six-foot nylon strap and buckle supplied with the camera (Figure 9.1*A*). Second, they can be cinched securely to tree trunks using a locking steel cable that fits 5⁄16-inch channels in cameras and steel cases (Figure 9.1*B*). Third, screw mounts suffice when no trees are suitable for strap or cable mounts (Figure 9.1*C*). Most cameras have a ¼-inch-20 threaded receptacle for attachment to tripod heads or to a variety of commercial screw mounts. DIYers make a simple adjustable mount from an eyebolt and a matching eye screw.[1] These are easy to bore into conifers, but you may need an awl or drill to start the hole in hardwoods. A crank-arm version of this design is easier to screw into wood and can be used with a variety of commercial or homemade mounts (Figure 9.2). A large lag screw will stay put better than small commercial mounts.

Older-model trail cameras lacked LCD viewfinders, so framing scenes was a matter of guesswork. Recent models have LCD viewfinders on the front of the camera, which allow you to view the scene and adjust the FOV. But the proximity of lens and viewfinder creates a problem: your face is in the way. Use an auto mechanic's mirror on an extendable handle to see the viewfinder without your mug in the picture (Figure 9.3). In bright light, however, the reflection on the LCD makes it almost impossible to view the scene. When this happens, trigger and view a few pictures to see if the FOV needs adjustment.

Figure 9.2. A DIY crank mount can be made from metal tubing with a welded lag screw. An inexpensive camera mount is clamped to this crank.

HOMEBREWED TRAIL CAMERA MOUNTS

Figure 9.3. An auto mechanic's mirror allows viewing the LCD when the camera is not in direct sunlight.

Homebrewed trail cameras can be mounted several ways. If they are fitted with eyebolts, you can attach cases to trees with bungee cords, rope, straps, or cables. They can also be attached to metal or wooden stakes (Figure 9.4). Cases with the **pipe-through** design can be attached with a cable lock or some cord. In either case, homebrewed cameras must be removed from the case to see the LCD and frame the photo.

The **Hickman mount** is a versatile adaptation of the pipe-through build that gives you access to the viewfinder.[2] Drill a **steel rebar pin** or a length of ¾-inch electrical conduit to accept a six-inch length of a ¼-inch-20 threaded rod (Figure 9.5). Tighten the camera on the horizontal rod with washers and a wing nut. Open the back of the case to see the viewfinder and tilt the camera or rotate the post to adjust the view. Round posts tend to rotate in loose soils because the camera is off-center, but you can prevent this by bolting or welding a flange to the stake. Stakes can also be lengthened using additional lengths of ¾-inch conduit and couplers available in hardware stores. Conduit is more portable than steel rebar pins and can be cut to fit in your backpack, but it's galvanized and sticks out like a sore thumb. A coat of drab paint will tone it down.

Interlocking mortis mounts are an alternative to cables, straps, and bungee cords (Figure 9.6). The slotted piece is screwed to the back of the camera case, and the mate is screwed to a tree, wooden stake, or metal post. They are easy to use and adjustable within limits when mounted on a post. Trail camera dealers sell additional mounting devices such as steel posts, clamp mounts for T-posts, and **gimbal** mounts that screw into trees or wooden posts. DIYers have machined their

A

B

C

Figure 9.4. ***A***, Eye bolts or u-bolts allow you to strap homebrewed point-and-shoot cameras to trees. ***B–C***, You can also bolt them to a crosspiece on a post.

Figure 9.5. The Hickman mount makes use of a threaded ¼-inch threaded rod and nuts to attach a pipe-through camera case to a steel rebar pin (concrete stake).

Figure 9.6A. A mortis mount made from aluminum channel allows quick attachment of the camera.

Figure 9.6*B*. A sliding tenon of scrap aluminum is riveted to a piece of square tubing.

Figure 9.7. A gimbal mount allows you to adjust the camera in horizontal and vertical planes. This one is made from sturdy sheet metal and square metal tubing. Welding is required. Tighten the bolts to fix the camera's position.

Figure 9.8. A quick-release tripod adaptor makes it easy to attach camera cases to a tripod. This large Pelican case has a mild steel plate screwed and glued to the bottom. The plate is tapped to receive the threaded tripod adaptor.

own gimbal mounts or made them from square metal tubing (Figure 9.7). If you're not a machinist, it's hard to make gimbals as functional as commercial products.

DSLR cameras in large cases require a tripod or other sturdy adjustable mounts, and a quick-release tripod adaptor is a convenient way to attach them (Figure 9.8). The quick-release plate attaches to the underside of the case and snaps into the clamp on the tripod head. DIY folk may choose to make a mounting plate from ¼-inch aluminum or cold rolled steel. Drill and tap the plate to receive the tripod's ¼-inch-20 screw, then attach it to the bottom of the case with steel pop rivets or small nuts and bolts.

Telescope window mounts made for birders and photographers are also convenient to use. Since the clamp opens only to about ⅝-inch, you will need an adaptor to fit the clamp to a post (Figure 9.9).

Ball-and-socket mounts are also versatile and easy to use. Small ones are inexpensive and suitable for lightweight PIR sensors and flashes (Figure 9.10), but large ball-and-socket **RAM mounts** are more reliable (Figure 9.11). The 1–1½-inch balls are capable of holding heavy cameras and flashes without slippage.

Figure 9.9. Telescope window mounts are easy to adjust. This one is clamped to a mounting plate on a post made of square pipe.

Figure 9.10. Inexpensive ball-and-socket mounts are suitable for lightweight flash units.

Figure 9.11. A RAM ball-and-socket mount connects a trail camera security case to a tree trunk.

MOUNTING POSTS

Materials for mounting posts are everywhere. You can buy materials in hardware stores or scrounge scrap pipe, aluminum, angle iron, rebar, T-posts, metal tubing, and PVC (Figure 9.12). Most prefer lightweight, portable posts. In bear, elk, and cattle country, T-posts or other heavy-duty posts are called for, and they must be deeply anchored. Square and round tubing comes in various materials, and different diameters can be used to make telescoping extensions or coupling sleeves to connect sections (Figure 9.13).

Whatever the post, you will need compatible mounts. To accept screw mounts, drill ¼-inch or ⁵⁄₁₆-inch holes in round or square metal posts or enlarge the holes in steel rebar pins (¾ x 24 inches). To accept a tripod head or window mount, make an adaptor. Though ungainly, heavy-duty T-posts usually hold their own against large mammals, but they always need an adaptor to accept even the most versatile camera mounts.

You can drive posts into soft soils by hand, but you'll need a rock or hand sledge to set a post in rocky or packed soils. A large rubber mallet is the ideal tool to prevent flaring metal posts intended to accept adaptors. A post driver is the best way to set T-posts, and a post puller is needed to extract them. I've resorted to these extreme measures to anchor cameras next to bear wallows. Round posts can easily be turned in the ground to adjust the camera's view, but large animals that blunder into them or rub on them can change the camera's view. Prevent rotation by welding or bolting a flange to the post, or buy commercial posts supplied with a flange or a pronged anchor.

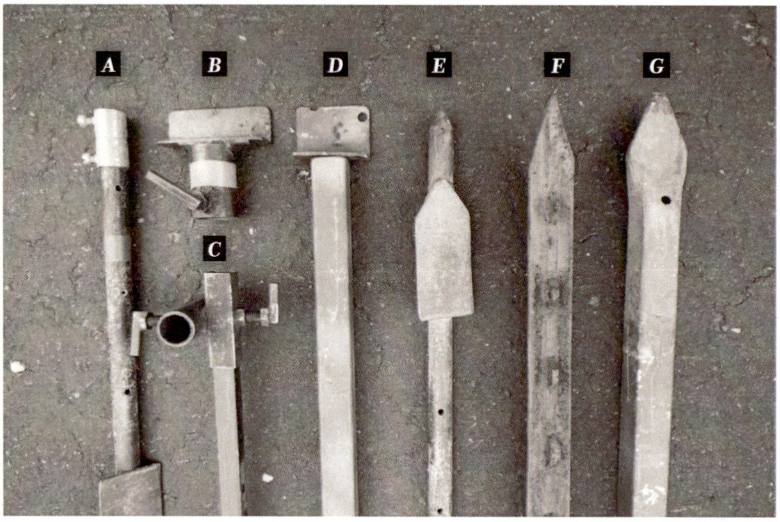

Figure 9.12. DIY metal posts and gadgets for staking cameras. ***A***, Steel rebar pin (concrete stake) with a coupler for conduit pipe extensions. ***B***, An angle iron mount welded to a pipe sleeve allows clamping a telescope window mount to a metal stake. ***C***, An adapter that allows you to connect a square cross arm to a conduit or a cement stake, as when mounting a flash to the cross bar (see also Figure 9.10). ***D***, Another angle iron mount than can be screwed into a nut welded into 1-inch square tubing. ***E***, A steel rebar pin with a welded flange to resist rotation. ***F***, A T-post ground to a point is easier to set in stony soil and is less likely to twist in the process. ***G***, This pointed tip was hammered, ground, and welded in a mounting post made of 1-inch square tubing.

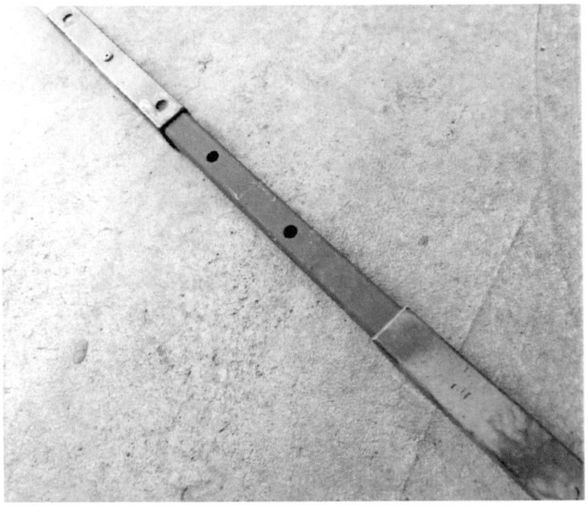

Figure 9.13. Lengths of ⅝-inch OD square tubing can be linked together using short sleeves of ¾-inch tubing. Set a pop rivet in the sleeve to create an internal stop. Hardware stores sell square tubing of various sizes. Discarded (rusty) square tube fence panels are also a source of material.

Figure 9.14*A*. A collapsible down and dirty camera skid mount with inexpensive fabric camo bunched up and glued to the box.
Courtesy Casey Smartt

Figure 9.14*B*. A rustic skid mount with three trail cameras for wide-angle coverage. The side posts pivot on lag screws for adjusting camera view.

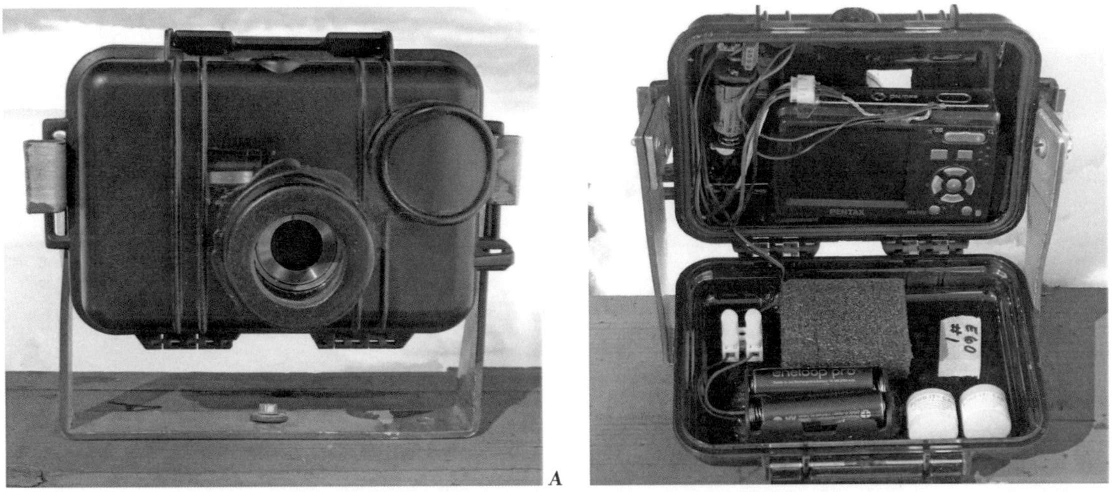

A ***B***

Figure 9.15. ***A***, An aluminum frame mount attaches to the strap loops of a Pelican 1020 case. ***B***, This mount makes it possible to open the case and see the viewfinder of this Pentax Optio E-60 camera.

Figure 9.16. A crane mount made from square tubing is another way to position a camera at nest cavities. Cinch the four-footed base to the tree trunk with a strap or a rope ratchet. A telescope window mount works best with this contraption.

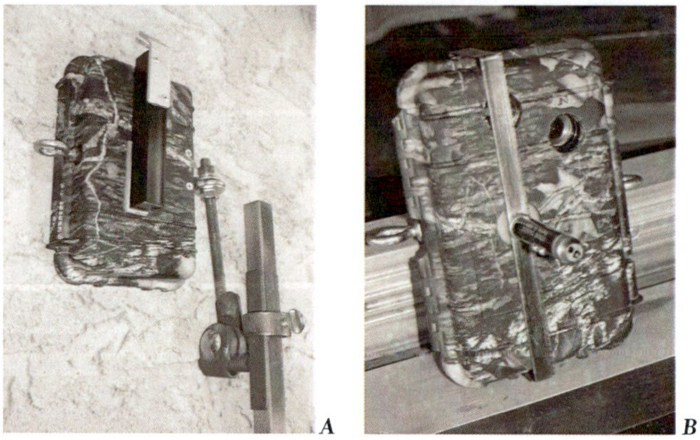

Figure 9.17. ***A***, A laser periscope fashioned from square tubing and a small piece of mirror can be attached to the camera with Velcro. Use the reflected beam to adjust the camera's position. ***B***, A laser sight can also help you aim a camera at an elevated tree cavity when the LCD is beyond reach. This one was made from a hack saw blade and a laser pointer with a magnetic base.

The "down-n-dirty camera skid" is a low-profile alternative for mounting DSLRs.[3] Attach your camera to the skid with a RAM ball-and-socket mount or a telescope window mount (Figure 9.14). Collapsible lightweight versions are portable. Rustic ones hammered together from 2 x 4s can be awkward to haul, but all are easy to deploy.

SPECIALIZED MOUNTING SOLUTIONS

Some camera trappers need specialized mounts in unusual locations or when seeking unusual perspectives. Safeguarding homebrewed cameras against curious bears also necessitates the use of security cases and special mounts. Special mounts generally arise from tinkering in the shop, and the resulting contraption may be so specialized it is useless for anything else. But if it makes a great photo or a new insight possible, who cares!

ALTERNATIVE LCD-FRIENDLY MOUNT

The LCD viewfinder is an important asset for framing images, but some homebrewed camera cases, like the Pelican 1020, are too small to admit the pipe for a Hickman mount. A workable alternative is an aluminum frame that fastens to the case's strap loops. I have used this for Pentax Optio E-60 cameras tightly built into very small Pelican 1020 cases (Figure 9.15).

CRANE MOUNT

Suppose you find a nest cavity in a tree but don't have a DSLR and a telephoto lens. If you can reach it with a ladder, you can use a **crane mount** to set a camera close to the cavity. This mount allows a range of camera positions; it is simply an arm with elbows that move in several planes (Figure 9.16). It's also friendly to birds: they will immediately adopt it as a perch and perhaps trigger a few blank images.

EXTENSION POLE AND LASER PERISCOPE

An extension pole is another aid to photographing elevated nest cavities when no adjacent tree is close enough to mount your camera (see chapter 7, "Arboreal Sets"). Telescoping fiberglass poles are made to fly flags and banners, but they can also "fly" a camera trap (chapter 7, Figure 7.17A) if you have perseverance.[4] First, you have to make a camera mount to fit the pole. Then, drive a T-post or a length of angle iron into the ground and attach the pole to it with ring clamps or cord lashings. Raise the camera to the height of the nest cavity and twist-tighten the sections to fix it in place. Stabilize it with guylines to minimize movement and resulting false triggers. Eyeballing an elevated camera's FOV from the ground is terribly deceptive, but you can aim it precisely when standing on a stepladder or orchard ladder.

A **laser periscope** is another aid to aiming an elevated camera (Figure 9.17). You can make one from a five-inch length of square tubing (at least one-inch diameter). Cut away three sides of the tube about 1½ inches from one end, bend the remaining piece to 45 degrees, and glue a small mirror to it. Velcro the periscope to the front of the camera case next to the lens, and then adjust the camera's position on the pole till the reflected dot is right next to the nest hole.

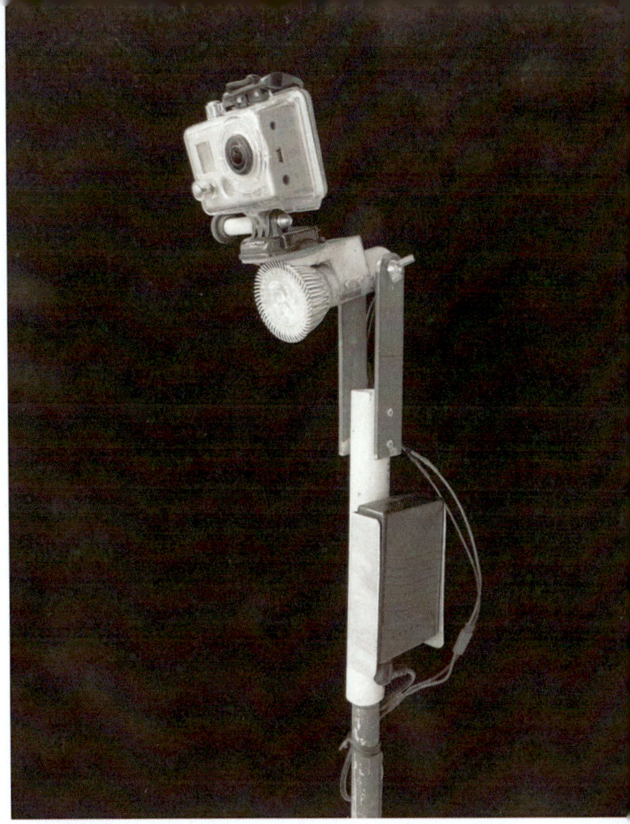

Figure 9.18. Make a pole camera to record video of elevated tree cavities or hollow logs. Switch on the power bank to the 9-watt LED lamp, power your camera on, and raise the pole to the cavity. The angle of camera and light can be adjusted with the wing nut. To take still pictures of elevated cavities use your camera's self-timer, but you had better be quick.

A B C

Figure 9.19. Camouflaged security cases made from metal conduit boxes. **A**, A case camouflaged with tape. ***B***, A mottled case painted with a sponge. **C**, A case sculpted with Bondo and painted for camouflage.
A: *Courtesy of Gary Schroyen*; **C**: *Courtesy of Woody Meristem*

A

B

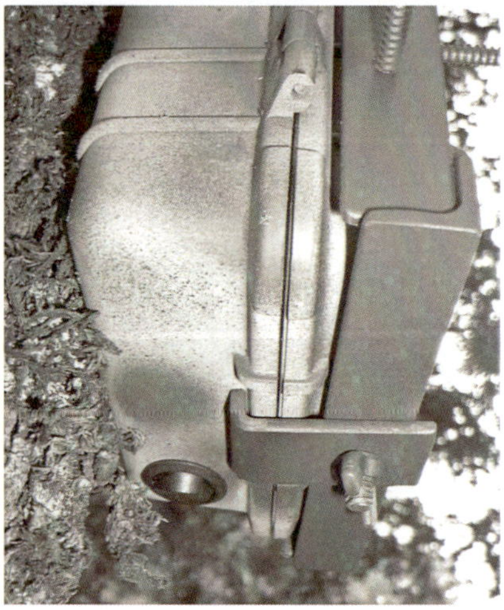

C

Figure 9.20. **A**, A bear guard that clamps to the front of a Pelican 1060 case made from angle iron and sheet metal screws. ***B***, The rigid top clamp grips the ridge of the case. ***C***, The bottom clamp can be tightened with the wing nut.

Aiming the camera with a laser periscope is a job for two. Point the laser up into the tube while your helper looks for the reflected laser point and adjusts the camera's position until it is right next to the cavity. Be advised: **parallax** can be a problem. When I moved my homebrew camera closer to a screech owl's nest cavity, the number of photos dropped abruptly. The camera's lens and motion sensor were several inches apart, and the screech owl's flight path was above the PIR's small field of detection. Parallax is not an issue with commercial trail cameras where lens and PIR sensor are close together.

Figure 9.21. Chain or cable can prevent a predator from stealing a game caller.

POLE CAMERA

If you fancy scoping out hollow logs and tree cavities, it's easy to make a pole camera. Figure 9.18 shows how I attached a GoPro Hero2 with a LED lamp on an adjustable arm. My pole of ¾-inch conduit measures 12 feet. At arm's length, it can reach 20 feet. Shorter versions work well with hollow logs on the ground.

Figure 9.22. This PVC scent delivery tube contains a wad of scent-saturated cotton. It protects the scent from rain and can disperse scent farther when in an elevated position.

EXPANSION POSTS

Expansion posts are helpful when setting a camera between solid stone surfaces, such as small lava tubes, mines, caves, recesses in rimrock, and between boulders in talus slopes. They may also help when setting cameras in large tree cavities. Use telescoping sections of round or square metal pipe to make an expansion post, and drill holes through both members at short precise intervals to accept a set bolt. The bottom should be capped with a flat edge to prevent turning. The short top section turns on a ⅜-inch threaded rod and has a pointed cap. After setting the maximum length with the set bolt, anchor the post tightly by expanding the terminal section. Attach the camera to the post with a handlebar RAM mount, adjustable clamp, or U-bolt.

Figure 9.23. ***A***, A utility box will keep an SLA battery dry; this one is vented on the bottom and shaded by a 5-watt solar panel. ***B***, A metal ammo case can secure a battery from most black bears. The rope is looped through single chain links on 4.5-inch lag screws. The cable between battery and camera should be hidden and made inaccessible to bruin.

OTHER ACCESSORIES

ALARMS AND DISTRACTIONS

Alarms are not common camera trap accessories, but flashing-light alarms are proven to deter some predators.[5] In theory, sound and light alarms could discourage wildlife and people from tampering with cameras. I was totally unnerved when I triggered loud beepers in a remote Alaskan mining claim. That spooky sound echoed across the bleak landscape from several directions, and all I knew was that I wanted out of there. Perhaps such an alarm would rattle would-be camera thieves before they encountered a camera trap. Just a thought.

Figure 9.24. I made this three-sided housing with a sheet metal hood to protect the camera from direct sunlight, driving rain, and snow in the Sierra Nevada.

SECURITY CASES AND BEAR GUARDS

Metal security cases are good investments against thievery and animal damage. They are readily available for commercial trail cameras. Metal conduit boxes, which come in various sizes, make good security cases for smaller homebrewed cameras (Figure 9.19). Heavy-duty plastic cases can protect DSLR cameras against snoopy bears, and a metal hood or sunshade may divert their attention from the camera lens. Watcher cameras reveal that the bruin treats spiked bear guards delicately (Figure 9.20). Spiders will anchor webs to spikes that point beyond the front of the case and create interesting but undesired special effects. Dress the spikes with Vaseline or tanglefoot as a possible deterrent to web-building.

ARTIFICIAL ROCKS

Cameras and accessories like large 12-volt batteries can be hidden in fiberglass rocks, which are sold in garden supply stores. Fake rocks that match rocks in a natural setting may succeed in hiding equipment from people, but they will not fool the bears.

GAME CALLERS

Commercial and DIY game callers are commonly used predator attractants, but if they lack protective housings and are not secured with chain or cable, predators may chew them up or abscond with them (Figure 9.21). (See chapter 8 for more about predator callers.)

SCENT HOLDERS

A short section of one-inch PVC wired to a tree branch is a handy way to disperse scent lures (Figure 9.22) because breezes spread scent much farther from elevated positions (see chapter 8).

Figure 9.25. Give your camera an "eyebrow" for rainy days by gluing a wedge of PVC pipe above the lens window.

SOLAR CHARGERS AND EXTERNAL BATTERIES

Longstanding camera traps can be powered for many months with supplemental C and D alkaline cells or Energizer Ultimate Lithium batteries, but sealed lead acid (SLA) batteries have more amp hours and last even longer. Commercial trail cameras have a 12-volt DC port for connecting SLAs, and it's a good idea to protect the battery in a utility or ammo box (Figure 9.23). Solar panels can charge SLA batteries and keep cameras running even longer in sunny areas when deep snow and inclement weather bar human access. Small adjustable solar panels are now standard features of some commercial trail cameras, and other models offer large accessory panels for charging 6–12-volt supplemental batteries. Homebrewers occasionally use high-wattage solar panels to power cameras for remote winter wildlife sets and long-term time-lapse photography. A word of caution: rodents like to chew electrical cords. Play it safe and use armored electrical cable between cameras and external batteries. Armored trail camera cables are now available on the internet; flexible PVC tubing is a cheaper but less effective alternative.

A

B

Figure 9.26. ***A***, A camera security box can be bolted to a sapling post using a U-clamp made from mild steel. ***B***, A camera bolted to a wood stake that was prepped in the field (a trail obstacle set).

OFF-CAMERA VIDEO LIGHTS

Several choices of off-camera LED lights are available for night video. You can also hack commercial video lights to interface with camera controllers, but LED floodlights designed for off-road vehicles are a more economical alternative. An 18-watt pod light can be powered for two to three hours using four 18650 batteries. The unit can be neatly built into a Pelican 1020 case with space for an additional set of batteries to double the amp hours (chapter 6, Figure 6.11). They can be triggered using electrical cables or with wireless RF transmitters.

SUN AND RAIN PROTECTION

Camera manufacturers caution us against the effects of high temperatures. Direct sunlight can heat and damage trail cameras, reduce battery life, and cause false triggers. In my experience, trail cameras tolerate high temperatures well, but a homemade hood or sunshade can limit direct sunlight and heating (Figure 9.24). Hoods can also protect cameras from whitewashing by perching birds, especially in treeless habitats. Camera hoods are insurance against leaky seals in homebrewed camera cases and are a safeguard where dripping water, ice, or snow on the lens window can ruin many shots. An inverted quart or gallon plastic bottle can weatherproof an electronic flash. Reinforce the screw lid to connect a post mount and a hot shoe. It will stand up to rain well. Bear resistant flash housings can be made of PVC or ABS pipe with a cap on one end and clear plexiglass on the other. Where bears are not a problem you can bag your flashes in clear plastic Ziploc bags and anchor them with rubber bands or cord.

Commercial wetting agents such as Rain-X reduce droplet formation on glass lens windows but must be refreshed when rain is frequent. "Camera eyebrows" may spare the lens from water drops during light rain (Figure 9.25). Heavy rain often splashes soil and plant material on cameras mounted close to the ground. You can reduce this annoyance by leaning a large piece of tree bark at an angle against the camera post. It will deflect raindrops away from the lens.

WOODCRAFT SOLUTIONS

Sometimes crafting mounting posts in the woods is preferable to hauling T-posts or metal stakes. With a pruning saw and hatchet, it takes only a few minutes to prep a mounting post from dead wood or to strip a conifer sapling and hew a point on the end. If you have drilled the camera case to receive a U-clamp, just clamp the case to the post. Alternatively bore holes in the post to receive the camera's metal security case and attach it with hex head screws and a nut driver (Figure 9.26). Set the post lightly with a rock or a rubber mallet, and then frame your picture. Remove the camera and drive the stake securely in place and then make final adjustments.

Homebrewed cameras may require more work to mount rustically, but pipe-through builds and eyebolts can be lashed to a cross piece with some cord. This is not a minimalist's solution. In addition to a pruning saw, rubber mallet, and axe or machete, you'll also need a knife, parachute cord, nut driver, and duct tape for good measure. But if you fancy yourself as a woodsy person, this quirky approach might just work for you. And sometimes, there may be no other choice.

CHAPTER 10

Camera Trap Security

Trail cameras are subject to a variety of harmless, often curious, and occasionally destructive reactions of wild animals and people. A metal security box will minimize the effects of rough treatment by animals. You can reduce the chance of human discovery and theft by setting camouflaged cameras with no-glow lights far from the beaten track. Cable locks and signs may deter some thieves. Hidden surveillance cameras, traceable SD cards, and tracking applications can help to locate stolen cameras. Mother Nature's whimsy can be blamed for some camera loss, but human error accounts for most cameras lost in the line of duty and is often due to failure to ask the *what ifs*.

It was a delightful find! Red tree voles usually nest high in Douglas firs, out of reach to all bipeds except arborists. This nest was only 15 feet above the ground. My host had kindly enlisted staff of the logging company to find it, but the sapling that contained it leaned toward the logging road from a steep bank. I tree hugged it with one leg and stretched to aim the camera at a limb leading to a sprig of short green needles. The tree had many such branches, but with luck the reclusive vole might descend from its nest, traverse this particular limb, and gather sprigs of its resinous fodder. With a little more luck, the camera might even shutter the critter hauling the cuttings home. It would be a dandy picture, but it didn't happen. My host emailed me a few weeks later and reported with some embarrassment that a masticator had eaten the camera (Figure 10.1). In case you are wondering, a masticator

Figure 10.1. The remains of my camera trap were recovered from the masticator's scat pile, which was chock full of wood chips and mulched greenery.

is a logging machine that seizes small trees, branches, and brush with an extendable arm equipped with a steel mulching apparatus. It's basically a mechanized brontosaurus that dumps piles of processed vegetation like fresh dung on the spot.

Cameras entrusted to nature are never safe from acts of God and mortals, but three painful lessons have made camera trappers wiser people. First is the realization that forgetting where you left your camera can be an expensive lesson. Second is the discovery that forces of nature, animals, and mean-spirited people can and will destroy cameras left in the woods. And third, when memory doesn't fail, there is the shock of finding one's camera just plain missing in action.

This chapter shares some camera trapping experiences related to security and describes measures that could stop the same thing from happening to you.

PROBLEM #1: FORGETTING WHERE YOU LEFT IT

Hiding cameras in the woods can have embarrassing consequences if you don't know the area, forgot the location, or didn't take a GPS waypoint. Seasonal change is one of Mother Nature's more kindly tricks because nothing in the woods looks like it did a few months earlier. Searching for a camera can become an ordeal of worry, but camera trapping forums have taught us that lost camera stories sometimes have happy endings.

A camera trapper in Texas wrote that he was out looking for his cows when he stumbled upon a camera cabled to a tree. He didn't remember putting it there, but sure enough it was his. A trail camera devotee in Upper Michigan, intent on setting a camera at a favorite location, was dumbfounded when he arrived and found one of his own cameras already there. He had set it the previous winter and then forgotten about it. Perhaps the record for lost-and-found cameras is a Wisconsin camera trapper who stumbled upon a camera he had set three years earlier. He had written it off as lost in action, but in a flashback he remembered moving it there in frustration with the lack of pictures at the previous location.[1]

To avoid such embarrassments, take notes on the route and location of the set, and photograph the set and its surroundings. Strategically placed lengths of surveyor's flagging are good navigational aids, but don't blaze your entire path and betray your camera's location to a stranger. With a GPS, you can record your path to the set using the track function; a proximity waypoint will sound off when you are near. If you have a camera with GPS capability, take a photo of the location and record a waypoint at the same time. You will still need a GPS unit to find the location. An electronic security token could be used to locate a trail camera. These sound-producing locator devices are as convenient to use as your car's key fob, but receiver–transmitter distance is presently limited to about 30 feet.

Figure 10.2*A*. Opening a drowned camera at the Sacramento Wildlife Refuge.
Courtesy of Ken Hickman

B

C

Figures 10.2*B*–*C*. It was a promising set until the tide came in. (Glenn County, California; Sony s600 IR conversion)

TAKING A CHANCE ON THE WEATHER

We set a trail camera on the edge of a swamp in the Sacramento Wildlife Refuge in December 2010. Flotillas of waterfowl covered the marshes; beaver and otter sign were abundant, and I hoped the camera was above the inevitable flood. I was wrong. Over the next six months about nine inches of rain fell in the area, and in May we found the camera underwater. The SD card told the sad story. The night we drove home, otters and raccoons left a dozen images during a gentle shower. Over the next four days more raccoons, black-tailed deer, and an egret visited. Though it had rained only an inch and a half, the tide started to rise. The camera recorded the rainfall at a rate of four inches in three minutes. The camera finally went under, snapping fifty photos of murky water before shorting out. It was an expensive lesson (Figure 10.2).

TRUST THE EXPERTS

Knowledge of local hydrology equates to an ounce of prevention. I asked a wildlife biologist if my cameras would be safe from floodwater along a creek in California's northwest coast. It was September, and rains could start any time. He told me that the heavily timbered catchment was on deep sandy soil and that it would take two weeks of steady rain to saturate it before the water would start to rise. I was probably safe for another month. Hydrologists are well aware of the factors that influence flooding: depth of permeable soil, soil saturation, rainfall, and vegetative ground cover. He was right, and I was impressed.

PROBLEM #2: DESTRUCTION BY WATER

Water is the most common threat to camera traps. High humidity can short a circuit and cause a camera to shoot repeatedly; wet climates will eventually corrode wiring. Putting packets of silica gel inside the camera case is the usual solution, but in humid climates silica gel quickly saturates even in a well-sealed case and must be changed regularly. When cameras are left outdoors for long periods, it's a good idea to dry them out on the mantle above a stove or near an electric heater or heating vent. Where this isn't possible, you can always "cook" a damp camera in the sun. Don't be alarmed when condensation shows up in the LCD or viewfinder; it will eventually evaporate.

Rainwater can enter homebrewed trail cameras through a faulty water seal in the housing, through lens and sensor windows, and through the holes drilled in the case for electrical jacks and external switches. Dirt can breach a good rubber seal, and even good seals eventually degrade and admit rainwater. Some seal the lid joint with a tight-fitting rubber band made from an automobile inner tube. Adhesive tape may also protect the seal from leakage, and an improvised camera shelter or hood may do the same, but these are temporary fixes.

Replacing leaky seals is the best solution. Test the rubber seal by immersing the empty camera case in a bucket of water for a minute or longer. Adjusting or replacing the seal usually corrects the problem. You can also seal a gasket with silicone, though it is messy and may smudge the lens. Camera trappers who deploy cameras for months at a time may opt to hermetically seal the camera in an aluminum case with a contact adhesive.

Wetlands are rewarding areas to photograph aquatic mammals and waterfowl, but if high water doesn't baptize your camera, men and boys with fishing rods may regard its discovery as a gift from God. Flooding can take place even where there is no precipitation. High water is seasonally predictable in undammed rivers. The ebb and flow of a coastal marsh depends on the tides; high tides can raise the water level many miles from a river's mouth. River flow in the mountains ebbs in the morning and rises in the afternoon, reflecting hourly changes in melting snowpack. Snowmelt has dramatic delayed effects in faraway places. In the Pacific Northwest, spring rain melts the snowpack and causes floods in coastal valleys. "Taking a Chance on the Weather" is an object lesson about precipitation and rising water. To learn about local watersheds check with resident outdoors folk and land managers employed by timber companies and federal land agencies. Some of them will be able to offer sound advice (see "Trust the Experts").

Many rivers have reservoirs where government agencies manage outflow to meet human and ecological needs. Discharge schedules for states and counties may be found online using the search term "reservoir release schedule." Dates and times are listed by reservoir, often with a caveat that "dates are tentative and subject to change *at any time* to meet power demands." The chance of losing a riverbank camera to reservoir releases may seem slight if you set it for a short time. Do your homework when in doubt. If you are working dammed rivers, search reservoir release schedules. If you are working wild creeks and rivers, consult the searchable links in "Searchable Maps . . ." for real-time data on river levels and flow forecast.[2] These precautions may prevent an unpleasant surprise.

SEARCHABLE MAPS OF FLOOD AND HIGH-WATER CONDITIONS IN THE UNITED STATES AND CANADA

- **USGS WaterWatch:** https://waterwatch.usgs.gov/?id=ww_flood
- **National Oceanic and Atmospheric Administration, National Weather Service River Observations:** https://water.weather.gov/ahps/
- **Wateroffice Canada:** https://wateroffice.ec.gc.ca/index_e.html

PROBLEM #3: DESTRUCTION BY MAMMALS

Finding your camera may be the silver lining in stories of lost cameras, but finding it mangled on the ground can stir vindictive passion followed by resignation and reflection. Camera destruction is one of the risks of camera trapping in bear country. Some accept their loss with resignation. Others vow to even the score come hunting season. Most consider it a rite of passage, chalk it up to experience, and take measures to prevent it in the future.

You are lucky if the destruction is incomplete. Canadian Arthur Veitch had set his DSLR at an elk calf killed by a cougar. He wrote:

> I found the cougar kill cached in a gully. I set up on it, leery of a grizzly procuring the carcass. A few days later, I approached the site from the ridge. From my vantage point, I saw the mound of debris where the kill had been, tripod legs and gear all mixed with the duff the bear had dredged up. Not wanting to pick a fight I couldn't win, I left and reclaimed my gear two weeks later. It appeared the bear had simply hauled the remains from the mound and reburied it a few meters away. Had it dug the elk calf up to continue feeding I think the damage would have been more extensive.

Veitch exhumed the casualty and restored it to working order.[3] Even in worst-case scenarios, a homebrew may shoot again if you salvage and replace parts. Ironically many bear-trashed cameras contain a parting gift that stirs irritation and wonderment—surreal close-ups of the culprit having its way with your dear camera.

Members of the dog family are normally shy around trail cameras, but some deviate from the norm. Jeremi Skelton found that a coyote had removed both his cameras and slave flash from their tree mounts and carried them away 50–60 feet. He recovered one camera and the slave flash, but the second camera remained under snow until spring. Tooth marks were the only damage.

Some wolf packs also have bold members.[4] Neal Wight was radio-tracking a pack of wolves in the Flathead Valley of Northwest Montana when they abruptly left their den to begin a sortie. He wrote:

> I found where they crossed the Flathead river . . . and eventually on an old logging road I found recent pup and adult tracks in the snow, mixed with regurgitated prey and tail-wagging sign. I set my Sony digital camera in a custom-made case and headed home. I came back the next day to find that something had opened the camera case, removed the camera and neatly clipped all of the connecting wires. The culprit, an adult wolf, was captured in a single photo before he destroyed my cam. Luckily, I was able to re-wire the cam, and it is still working.[5]

Small animals can trash trail cameras too. Rodents evidently find plastics particularly chewable and are known to gnaw camera cases and insulation on electric and security cables. In Pennsylvania, Woody Meristem had a porcupine gnaw a hole in a plastic Fresnel lens and had gray squirrels gnaw camera cases that were sculpted with "3D camo" using Bondo, a synthetic auto-body compound. "The squirrels took off most of the Bondo and were working on the case—and probably would have gotten through it if it had been left out for two more weeks."[6] Tree felling is a minor but real liability in beaver country. My friend Brian Miller found no trace of his camera when a beaver felled the cottonwood tree it was on in New Mexico. The rodent had cut and hauled away most of the tree, and he concluded that the camera was now part of the beaver's dam. He accepted his loss with humble grace.[7]

LESSER IRRITATIONS

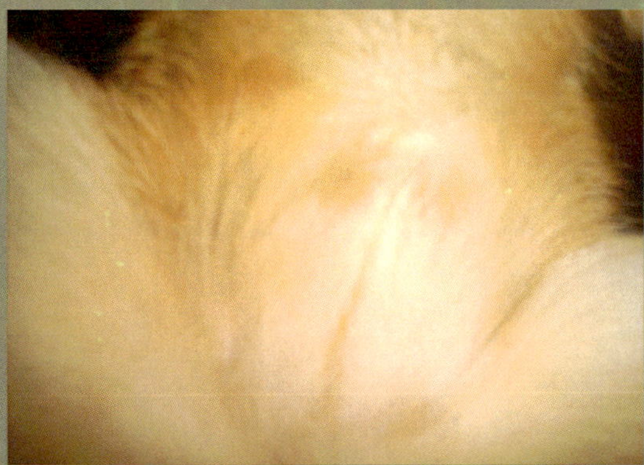

Figure 10.3. Caught in the act: A bear cuffed the camera to the ground, and then this cheeky gray fox squatted and urinated on it. (Kern County, California; Sony s600) *Courtesy of Ken Hickman*

Meddling with trail cameras by mammals and birds may amuse or irritate the camera trapper. Some elk and cattle are ardent lickers of cameras, and their saliva transforms the lens window into a soft-focus filter. Cattle, feral pigs, and bears have used remote cameras as scratching posts, which almost always changes the camera's FOV. Elk that use cameras as sparring partners may topple the camera and its post, as once happened to me. I found the camera embedded in mud and covered with trampled vegetation. California nature blogger Ken Hickman wrote about a bear that smacked his camera off its mount. Fortuitously, this switched the camera shoot mode from night to day, and the camera captured sixty-one pictures of the curious offender examining the device in broad daylight. The camera was none the worse for wear until a gray fox urinated on it eleven days later and at least twice more over the next two weeks (Figure 10.3). Hickman found a sticky, stinky camera, but it was still "working like a champ, taking photos."

Figure 10.4. Casualties of curious bears: **A**, The bear removed the Fresnel lens despite the "protective" spikes on a metal frame. ***B***, A yearling toppled the camera and shattered the LED diffuser inside a hollow log. (I failed to remove the corks from the spikes.) **C**, This remote RF motion detector performed well until a bear deconstructed the aluminum mount and chewed the Fresnel lens. ***D***, The PVC lighting poles snapped under duress; the sow only wanted to sniff the LED arrays.

Predators have occasionally absconded with equipment too, including cameras, decoys, and callers, but victimized camera trappers often find the missing equipment nearby.[8]

In tropical Asia, wild elephants are the camera trapper's nemesis. They may only examine the camera by sniffing and fondling it with their trunk, but often they seize it in a steely grip and twist it from its mount. Sometimes they finish their inspection by stepping on it.[9] Fortunately, the attentions of some animals are only lesser irritations (see "Lesser Irritations").[10]

A metal housing is your camera's best bear deterrent. Lock boxes or security boxes can be purchased for many models of commercial trail cameras, and homebrewers can modify several kinds of metal utility boxes as protective cases. Those with the time and tools may build protective cases from sheet metal. Another deterrent is a spiked angle iron frame fitted to the camera case. Bears make contact cautiously with strange objects and may desist when they detect the spikes, but not always (Figure 10.4). The disadvantage of spiked frames is that they attract spiders, which anchor webs to the spikes and cross the camera's FOV. Spider webs make disconcerting visual effects in the glare of the flash. Chapter 9 discusses construction of these devices.

PROBLEM #4: DESTRUCTION BY INSECTS AND SPIDERS

Arthropods are the smallest assailants of trail cameras.[11] Several ant species build nests in manufactured structures such as utility boxes, bird boxes, and trail cameras, and with the establishment of a queen the nest becomes a perpetual ant factory. Circuit boards and batteries apparently give off scents that attract ants, which can short circuits and trigger the shutter repeatedly. Cameras showered with honeydew from trees or contaminated with traces of bait also attract insects, but ants don't require such attractants to set up housekeeping. Trail cameras with small openings are simply good places for ants to nest.

There are several ways to eradicate camera-nesting ants. If you have the inclination and patience, apply commercial ant bait and wait for the workers to poison the queen. This can take several days to over a week. Spraying with insecticide is another option, but not recommended. Poisons that kill ants on contact often don't reach the queen, and insecticides contaminate the camera and also poison the soil. Sealing the camera in a plastic bag and putting it in a freezer for a couple days will kill the colony. This works well with less aggressive small species. Shake loose dead ants or blow them out with a can of compressed air. You might also try contacting the camera manufacturer to explore the possibility of getting it cleaned. (Good luck). The introduced fire ant (*Solenopsis invicta*) is an aggressive species whose bites are painful, but fortunately it is not a verified encroacher of cameras. Should a camera become infested with fire ants, treat it with an ant bait product and leave it alone until the colony dies. Mothballs and pieces of plastic pest strips are reported to be effective repellents; but because they contain toxins, handle them with care and avoid the fumes.[12]

Paper wasps, hole-nesting wasps, solitary bees, and caterpillars may also move into homebrew cameras, but this is uncommon. Paper wasps can sting, but the main offense of hole-nesting wasps and bees is plugging holes in the camera case with mud or leaves. The obvious prevention is to seal holes; the deterrent to spiders on spiked frames is to apply Vaseline or tanglefoot on the spikes.

Tropical termites can wreak havoc on cameras lacking protective housing. In the Peruvian Amazon, biologist Jeff Cremer found that termites had taken up residence in his DSLR camera. These particular termites, a species of the pan-tropical genus *Nasutitermes*, are known to bore through PVC pipe in Australia. They use their feces to construct galleries and passageways. Cremer's camera, encrusted and packed with termite feces, was beyond repair.[13] Sealing your camera in an insect- and weather-proof housing is good practice anywhere.

PROBLEM #5: HUMAN VANDALISM AND THEFT

There are three levels of defense against human vandalism and theft. The first is cryptic behavior: be cagey when placing your camera to reduce the chance of someone finding it. The second line of defense comes into play when someone finds your camera. In this situation, deceptive tactics and physical deterrents may discourage the would-be thief. The third line of defense includes tactics that can lead to the identification and apprehension of the successful thief. We will look at each of these strategies below.

REDUCING THE CHANCE OF SOMEONE FINDING YOUR CAMERA

Though there are kind and benign camera finders (see the prior sidebar), other finders have trashed trail cameras when they realized they have been photographed, and would-be thieves have vandalized cameras when the locked steel cable thwarted their efforts to steal them. Patient thieves have returned with bolt cutter or chain saw to finish the job. Such experiences have transformed normally lovable camera trappers into brooding fiends who subject their families and friends to tiresome reruns of imaginary Stephen King–like scenarios of revenge.

It seems that most camera finders endorse the finders–keepers philosophy. Camera trappers agree that if a thief really wants your trail camera, they'll find a way to get it. It makes sense therefore to reduce the chances of people finding it. So let's consider ways.

KIND AND BENIGN CAMERA FINDERS

I'll never forget the photo-capture of a gloved hand with pruning shears trimming vegetation in the camera's way. It was the hand of a kind and unknown camera finder. Cameras make people do strange things. I found a long, handwritten note in a commercial camera I had staked on a friend's property. It was signed by "Da Bear." It proclaimed the imposter's paternal concern for each of the species on the SD card. Obviously, Da Bear had removed the card, viewed the photos, and composed its prank before returning it to the camera. The single video file that was missing was probably the one that recorded the droll pretender's identity.

Figure 10.5. Though my camera innocently trolled the hikers repeatedly, its payback was a surprise encounter with menacing animals.

Benign finders of trail cameras have paused for a selfie or a wave and a smile. One of my water sets on a ranch recorded a boy smiling at the camera while urinating in the creek. At a remote long-term set I was alarmed to find clips of a shirtless adolescent apparently prospecting with a new rock hammer. When he saw the camera he approached, inspected it like Sherlock, and mused, "Hmm, Cal Fish and Game!"

On a ranch in the northern Sierra, the head of a small boy appeared suddenly in the frame, upside down. Several weeks later, the boy again popped into view, glared at the camera, and beat his chest like a silverback gorilla. Then he hurried off to follow the leader, a winsome, brown-eyed girl.

School groups encountered my cameras regularly at the Audubon Canyon Ranch, and they mugged like monkeys when they found they could trigger the camera to take flash pictures. When the teacher herded her flock away the camera captured the last picture—a blond, blue-eyed cherub coyly giving my camera the Anglo-Saxon mono-digital display.

A camera on one of the back trails became the bane of some lady volunteers. It was set for **red-eye**, so the flash made three weak but quick eye-catching bursts before the big flash exposed the picture. It was as if a troll had jumped out of the ferns and flashed the ladies. Their endearing gotcha moments changed from portraits of shock and dismay to a certain eye-rolling resignation, then finally to knowing smiles. I knew they had settled the score when I saw the photos of a toy skunk and chipmunk menacing the troll (Figure 10.5). These fine ladies kindly gave me permission to post this strange episode and photos in my blog. Maybe those gotcha moments were bad karma. The troll camera disappeared that winter, possibly stolen by mushroom poachers.

I sometimes wonder where my stolen cameras are now. Hidden in a box in someone's garage? On display in a pawn shop? In a landfill? Or perhaps resting in the Monterey Trench with other tokens of human folly?

SELECT A HIDDEN SITE

Set cameras off the beaten track where most people just don't go. You may have an intuitive feeling for such places, especially if you know the area well. Making a set within view of a highway, county road, firebreak, off-road vehicle track, or hiking trail is an invitation to thievery. Signs of human visitation, from cigarette butts to paint-blazed trees, are messages to keep moving. Equally good sites can be found away from trails, roads, and fly-casting sites, but finding them requires effort.

Select areas that escape the normal human line of sight. With the exception of birders, people don't often look up into trees or under bushes or low-lying vegetation. If you can't resist taking a chance, visit your set often. You will probably pull your camera if you are lucky enough to photograph a person before someone else steals it.

Camera trappers may feel hard pressed to find areas where a camera is safe during hunting season, when hunters flock to the outdoors from near and far, in darkness and daylight, and gain access to isolated areas with all-terrain vehicles. Small-game hunters roam widely, and large-game hunters scan landscapes from tree stands and lookouts. Visibility is often better during hunting seasons when deciduous trees and shrubs are leafless. Tracks in snow can give away the camera trapper's activities. White flash or red infrared array can alert strangers to a trail camera in predawn hours, and camera trappers who use carcass sets run the risk of attracting hunting dogs and their owners. Dense woods and brushlands are safer places to set cameras in hunting season; it's a good idea to wear blaze orange when you do it.

Geocaching is another recreation that draws people off trails and into remote areas. Geocachers use GPS units to search hidden locations for containers with logbooks and sometimes trading objects. To find out if it's popular in your area, search the internet for local geocaches. You may decide to avoid areas near waypoints in your county.

Finally, there is the problem of illegal marijuana farmers, our latter-day analog of moonshiners. Avoid their so-called grows, often encountered in remote sections of national forests and Bureau of Land Management (BLM) lands. Rather than stealing your camera, marijuana farmers are more likely to send you a hostile message by destroying it (see "Coyote Candy and a Pot Garden").

COYOTE CANDY AND A POT GARDEN

Napa County, California. What caught his eye was a zoological curiosity—a brightly marked candy wrapper of Bolitochas Tamarindos embedded in a coyote scat. Professor Richard Tenaza knew the area well. He often took his biology classes to the reserve for camera trapping and birding. Did the coyote pilfer the candy in a vineyard, he wondered. If so, it traveled at least a couple miles to this remote section of the inner coastal range called the Cleary Reserve. The curious find foreshadowed a series of events that led to a harrowing near-death experience.

A few weeks later the professor made a disturbing discovery. Several of his cameras were vandalized beyond repair. No bear would go to this extreme of destruction. It was a strong statement by someone who had an issue with trail cameras. He pulled his other intact cameras the same day. Not long after that the sheriff's deputies found the marijuana garden—six hundred plants hidden in the forest on a steep slope and watered with PVC tubing siphoning water from a spring upslope. The sheriff's narc squad removed it.

Then the professor and some zoological colleagues stumbled into an abandoned camp while scouting one of the reserve's backwaters. This is where the pot gardeners had cooked their meals on two propane burners, sheltered under a tarp, and evidently listened to CDs in their rustic beds—leaf mattresses framed with sticks. There were the candy wrappers he'd seen earlier. "I surmised that the mysterious candy wrapper had probably originated at the pot camp."[14]

The place had to be cleaned up, but a decent trail had to be blazed before his volunteers could bag the trash. For this task he enlisted Steve Anderson, a close colleague and fellow professor at the university. Despite a lingering uncertainty about the presence of pot gardeners, they felt safe bushwhacking in the woods, until late that afternoon. That's when Steve tripped and fell into a gully, gashing his head and fracturing a neck vertebra. Rich stanched the bleeding, and eased his friend onto a slightly level place, but there was no cell phone service in the ravine and no time to hike out to get rescuers before nightfall. If he abandoned his friend, the scent of blood might attract a cougar or bear. So the two aging biologists spent the night dozing and waiting for daybreak.

Steve's wife knew something was wrong when he didn't call home that evening, so she called the sheriff's office. The next morning a sheriff arrived at the reserve just as Rich reached the gate. He radioed for help and soon two helicopters, an ambulance, fire truck, and sixteen rescuers arrived. One helicopter stood by on the road while the other flew over the forest and hovered above where Steve was lying. EMTs slipped Steve into a tube, the hovering helicopter lowered a cable, and the professor watched his old buddy ascend.

It was several months before Steve recovered.

On trash pickup day, the professor drove to the reserve and hiked solo to meet the narc squad helicopter. Again there was no place to safely land. That's when he realized that this was also going to be a training exercise, more specifically, a mock pot garden bust. The pilot lowered each of three energized and armed narcotics trainees in different locations where they could be heard barging about and barking commands. "Eventually, they emerged from the bushes, and we got all the trash lifted up to the chopper."[15]

The experience gave new meaning to an aging cliché. You never know what to expect when camera trapping. When a local paper printed the story, an irate resident took issue with the rescue's expense to taxpayers. As wealthy as it is, Napa County is "no country for old men." At least old men who don't play golf. Thus ended a disturbing distraction in the usually benign pursuit of zoological knowledge.

Figure 10.6. Natural materials can hide a camera from human eyes. *Courtesy of Gary Schroyen*

CAMOUFLAGE

Animals have evolved color patterns that blend with the background. Natural camouflage is hard to see because it breaks up the outline and shape of the body. People have long used deceptive coloration for various purposes. Many trail camera users prefer their cameras painted in camo or even textured to resemble tree bark or vegetation, which is called 3D camouflage.

Trail camera camouflage is self-defeating, however, when the eye-catching lens, flash, and sensor resemble eyes and toothy grins. Homebrewers have overcome these deficiencies by putting camouflage tape and marker pens to shiny camera parts. Manufacturers could make camouflaged trail cameras that are more than designer statements. A dull finish, disruptive patterns on the Fresnel lens, and a disguised infrared array would be improvements.

Camouflaging a camera with bark, lichen, and dried leaves is an alternative to painted camouflage (Figure 10.6).[16] Fastening the materials with Velcro or glue is time-consuming, but for those with an esthetic bent it may be worth trying. Fiberglass and polyurethane rocks designed to hide security cameras are another option. Though often pricey, they can be customized as trail camera and external battery housings. Some have fashioned camera housings to look like an old stump. In the proper setting, artificial rocks and stumps can be effective disguises. One clever camera trapper reported hiding his camera in an abandoned hornet's nest. It lasted through the hunting season, apparently until someone blasted the nest with a shotgun.[17]

Cameras mounted on posts are highly visible in grasslands and deserts, but they can be easily hidden when staked next to a bush. Camouflage and props may disguise trail cameras visually, but they do nothing to disguise camera scent. Most mammals smell far better than they see, so when a camera is downwind, they easily detect it. We can fool people with camouflaged cameras, but not the mammals we seek to photograph. They will almost always find our cameras with their noses.

DETERRING THEFT WHEN YOUR CAMERA HAS BEEN FOUND

If a stranger finds your camera, the first lines of defense have failed. There is now a clear and present danger of tampering, theft, and destruction. The only hope of seeing your camera in one piece again is to thwart the person's criminal inclinations.

LOCKING THE CAMERA IN PLACE

Cinching the camera to a tree with a steel cable can discourage would-be pilferers. Several brands of lockable steel cables are commonly used, but even homemade cables can deter, though they are not as convenient to use. Using a cable on a homebrew camera requires the pipe-through design, in which a metal or PVC pipe penetrates the camera case. The case can't be opened when cabled to a tree through the pipe. A cable can also be looped through eyebolts on the side of the camera. Cables will not stop a thief from cutting it with a bolt cutter or a vandal from bashing the camera with a rock, but a metal lock box or security box will make it harder to damage. Trail camera sellers offer lock boxes as accessories, but padlocks will not deter the determined (Figure 10.7).

IDENTIFYING OWNERSHIP AND PURPOSE

Write or engrave your name, address, and telephone number on the inside of the camera case and keep a record of the camera's serial number. Knowing this number is proof of ownership if the camera is sold to a pawnshop or listed for sale on eBay or Craigslist.

CAMERAS WITH MESSAGES

A message on a camera is a mind game that may give second thoughts to those who find your camera and struggle with temptation.[18] Messages may be statements of fact or baseless warnings; a bold thief may take a chance and call the owner's bluff (see "Sample Messages to the Tempted"). Camera trappers who falsely warn of wireless communication had better supply the camera with a bogus antenna. A few years ago, a hunting forum member reported an incident recorded on his trail camera's video. Two would-be thieves were starting to steal the camera when one of them noticed the antenna on the camera and remarked that he thought it was streaming images to a cell phone or computer. The antenna was a fake, but the sight of it changed their minds.

SAMPLE MESSAGES TO THE TEMPTED

- Wildlife Survey Camera. Please do not disturb. Photos of people will be erased.
- Password protected. It won't work if you don't know the password.
- Anti-Theft GPS Tracking Chip
- If you can read this, your picture was just sent to my cell phone.
- GPS-enabled camera. Steal it and you will hear from the police.
- Do not disturb. Camera streams video to cell phone. Your identity has been recorded.
- This camera belongs to a veteran. Steal it and I will hunt you down and get even.

Figure 10.7. A distressing discovery. A thief has your camera. (Vancouver Island, British Columbia)
Courtesy of Gary Schroyen

FRUSTRATING AND IDENTIFYING THIEVES

My survey of camera trappers on the *Camtrapper.com* blog found that 34 percent of 234 respondents had their cameras stolen in the field. Victims of trail camera theft often give up hope, but 7.6 percent of them reclaimed their cameras. There are measures that give you a fighting chance to identify thieves and get your camera back, but statistics on their relative merits are hard to find.

PASSWORD PROTECTION

Some commercial trail cameras give owners the option of entering a password to access certain camera functions. Password protection denies thieves access to the camera's full functionality and prevents them from changing settings or viewing photos or videos. To what extent it deters theft is unknown. Thieves will not know that a camera has limited utility unless it is labeled "password protected." If you have a password-protected camera, enable protection and label the camera as such. It may deter some of the thieves.

HIDDEN SURVEILLANCE CAMERAS AND WIRELESS TRANSMISSION OF PHOTOS

If you want to photograph the thief in action, monitor your trail camera with a hidden camera nearby, preferably one with invisible infrared array. You might document the theft on video. Owners of wireless or cellular trail cameras can monitor photos and video clips remotely. A few years ago a trail camera owner in North Carolina received an email with a photograph of a young man in a truck smiling smugly. The owner had no idea his cameras were missing, but the photo told it all. Deputies sent the suspect's photo to local media, and numerous respondents provided information. Police arrested the thief and his accomplice.[19]

Trail cameras that lack built-in Wi-Fi can gain protection from SD memory cards with Wi-Fi capability that offer a means of tracking usage of the SD card itself. Such devices include Eye-Fi wireless SD cards and tracking applications such as iHound.[20] Eye-Fi cards upload photos to a home computer or website via the company's server. Thieves uploading photos from stolen cameras disclose their location, which is encoded in GPS-enabled cameras. Wi-Fi-enabled cards automatically tag photos with geographic coordinates. The iHound app alerts the owner when the SD card or trail camera is connected to an unknown computer. The Apple AirTag would be another means of tracking a camera thief, but its transmission range is only 800 feet.

SEARCH THE INTERNET FOR THE STOLEN CAMERA

StolenCameraFinder is a search service that scans the internet for photos uploaded to popular photo-sharing sites.[21] This is made possible by camera serial numbers usually recorded in EXIF data appended to each digital photo. StolenCameraFinder has a lost and found registry. Once photos from your camera are found, you can try to track down the thief using profiles on social media and photo servers.

GPS locater devices or GPS trackers also make it possible to track stolen goods and may have utility for camera security in the future. Apple's AirTag is one of them. They are small, contain a GPS receiver and a cell phone transmitter, and require a mobile phone service plan. Battery life varies depending on the product and range is a limiting factor.

CRAIGSLIST, EBAY, AND PAWNSHOPS

If someone tries to sell your stolen camera, Craigslist, eBay, and local pawnshops are logical places to look for it. An image of yourself on the memory card could help if a pawnbroker looks at the photos and sees you holding this message in large letters: "If the person who sold you this camera doesn't look like this person, the camera is stolen." Consult the police whether or not you are able to identify the thief who sold your camera.

INSURE YOUR GEAR

If previous options are too much trouble, you can always add your trail cameras and other expensive equipment to your homeowner's or renter's insurance. Make a photographic inventory of your cameras, and list serial numbers and purchase prices for your insurance company. You will still have to pay a deductible, but you will be able to replace your cameras.

I can't assure you that these preventive measures and tactics will work all the time, but certainly they can help to reduce loss of cameras resulting from lapses in memory, poor judgment, and theft.

CHAPTER 11

Ethics and Sensibility

> The process of photographing wildlife and the treatment and captioning of images are subjects of ethical concern. Camera trapping is a relatively benign form of recreation, but uninformed decisions can reinforce risky behavior and endanger wildlife. Biological knowledge and sensitivity to possible consequences of human action are also vital when photographing critical phases of an animal's annual cycle. This chapter discusses the ethical guidelines of organizations for nature photographers, truth and disclosure of photographic methods, the effects of wildlife attractants and artificial lighting, the concepts of human and animal privacy, and fair chase.

Pioneers of camera trapping touted their hobby as a humane alternative to hunting, but their use of explosive flash was an inescapable contradiction. Frank Chapman condemned the explosive flash as unethical. F. W. Champion thought it should be used with discretion, while George Shiras rigged his camera sets to capture the excitement of the flash's consequences. Each viewed his activity through a different ethical prism.

Camera trapping and wildlife photography are not necessarily innocuous.[1] Competition, lust for recognition, or meeting production deadlines can lure the camera user to cross the ethical borderline. Responsible wildlife photographers, however, are finely attuned to nature, act responsibly, and try to minimize their ecological footprint. Watching wildlife for thousands of hours garners a gratifying reward—a keen understanding of animal behavior.

The late Niko Tinbergen showed us how curious naturalists venture softly into the lives of wild animals to understand their minds and perceptual worlds. His gift for observation and experiments with free-ranging animals won him the Nobel Prize, and he remains an inspiring example to naturalists, and to camera trappers in particular.[2]

Viewers of wildlife photos also have diverse perspectives, and some don't always like what they see. Camera trappers may be praised or condemned for sharing photos on the internet. The critics may have an ethical axe to grind or just want an explanation. Wildlife film producer Chris Palmer wrote: "When viewers discover that something in a wildlife film causes unnecessary suffering or is packaged, inauthentic, or contrived, they feel cheated or misled."[3]

It is wise to be aware of public concerns about camera trapping. This chapter begins by reviewing online resources about the ethics of wildlife photography and then examines the ethical issues of using trail cameras and how to minimize risks to our subjects.

ONLINE SOURCES OF INFORMATION AND GUIDANCE

Organizations for nature photographers have identified two areas of ethical concern: the process of photographing wildlife and the treatment and captioning of images. The goal of the North American Nature Photography Association (NANPA) is to "promote thought and discussion among NANPA members and allow each individual to make informed choices about ethical behavior."[4] Its *Principles of Ethical Field Practices* advocates environmental sensitivity and conduct that protects the photographic subject and the photographer. It advises truth in captioning and disclosure of the subject as wild or captive. NANPA endorses standard printing treatments such as color correction as acceptable practice but cautions against representing digitally altered photos as reality. Cutting and pasting content from other sources into original photos is an unacceptable deception. In a nutshell, NANPA advises photographers to become knowledgeable of wildlife and ecosystems, to use common sense and obey laws, and to be tactful and courteous with others in the field. It also makes the case for being a good role model and educating novice nature photographers.

The Nature Group, a special interest group of the Royal Photographic Society based in the United Kingdom, also gives detailed guidelines on photographing plants and animals in its *Nature Photographers' Code of Practice*.[5] Its basic message is that wildlife photography can disturb animals and place them at risk; thus the subject's welfare takes priority to the photograph. Sound knowledge of natural history should guide the photographer's actions. Photographing nesting birds, for example, requires understanding of breeding behavior. A blind or hide may be necessary if the photographer's presence affects normal activities, but the structure should not attract the attention of the public. Song playbacks and decoys should not be used during the breeding season, of course, because they can disrupt time and energy budgets.

The International League of Conservation Photographers (ILCP) is dedicated "to further environmental and cultural conservation through ethical photography" and is guided by the core values of conservation, integrity, and photographic excellence.[6] Its policies focus on field practices, use of captive animals, digital manipulation, and truth in captioning. The welfare of photographic subjects, especially breeding animals, depends on awareness of "safe, informed, and responsible working distances." ILCP acknowledges the close ties between indigenous people and wildlife and advises respect and accurate representation of their values and traditions. Imagery of captive animals can play a significant role in promoting conservation of rare species, but ILCP cautions against patronizing for-profit organizations unless they practice high standards of husbandry.

These sites give sound guidance, but they do not deal explicitly with ethical issues specific to camera trapping.

INVADING HUMAN PRIVACY

Privacy, or "the desire of people to choose freely under what circumstances and to what extent they will expose themselves, their attitude, and their behavior to others," is considered a fundamental human right.[7] Laws grant citizens the right of privacy in various forms, including the inviolability of the home, secrecy of communications, and the control of personal information. Surveillance cameras have been called technologies of privacy invasion because CCTV was designed specifically to monitor human behavior. Americans agree that crime deterrence is for the public good, so most accept surveillance in public areas. Trail cameras occasionally photograph people on public land. Those who stumble upon a trail camera may find it an annoying breach of privacy, but the law does not protect their privacy in a public place.[8] Historian Etienne Benson wrote that users of remote wildlife cameras threaten the wilderness values of people who seek solitude, self-reliance, and "temporary escape from the everyday pressures of modern life." His question, "Can such values survive in a 'wilderness' blanketed with surveillance devices?" conjures a disturbing scenario of trail cameras populating the woods.[9]

The remote camera's potential to invade human privacy delayed the adoption of recreational camera trapping in Sweden, where the Right to Public Access law gives Swedes the freedom to roam the countryside just about anywhere on public and private land. Nowadays camera trapping for scientific purposes requires a government permit, and recreational camera trapping on private land requires written permission from the landowner. Camera trappers must post a sign announcing the presence of camera traps and provide personal contact information. To respect human privacy, cameras are not allowed on roads or trails, and accidental photos of people must be deleted immediately.[10]

Trail cameras are indiscriminate recorders of wildlife and people. A few years ago, when a camera left in a hunting preserve filmed a tryst between an Austrian politician and his mistress, the camera's dark side became newsworthy. Legal experts felt that the presence or location of remote cameras should at least be marked with signs so that forest visitors could "adjust their behavior and avoid the monitored areas."[11] Ironically perhaps, a few years earlier, a trail camera recorded a pair of Austrian thieves with stolen property. When they were arrested, no one protested their right to privacy.[12]

Trail camera users in the US are within their rights to post photos of suspected camera thieves on the internet, and Trailcamtrespasser.com (now expired) was created with this in mind. Respecting the privacy of the public will simplify your life as a camera trapper. There's little reason to keep images of strangers. You should have their written permission to sell their image, and that is rarely possible when using remote cameras.

Figure 11.1. One afternoon in March, a gray fox took a two-hour nap in a sandstone cave. The camera snapped seventy-seven pictures as the beautiful sleeper gravitated to the cave entrance. An invasion of privacy? The gray fox didn't seem to mind. ***A***, arrival (3:09 p.m.); ***B***, napping (3:10 p.m.); ***C***, still asleep (5:07 p.m.); and ***D***, wakeful and rested (5:08 p.m.). (Chimineas Ranch, San Luis Obispo County, California; Sony s600)

INVADING ANIMAL PRIVACY

Privacy has moral and legal implications for people, but no laws accord privacy rights to wildlife. Rights and privacy are concepts, not inherent qualities of life. Writer Jim Amrhein put it this way: "In the natural world, all that any creature—man included—has a 'right' to are those things it can take by force or forcibly defend from being taken."[13]

Predators violate the privacy of other species as a way of life. The solution to survival in nature is not a bill of rights but natural selection. Of course, we do accord rights of privacy to pets and wildlife when we feel it is deserved. The old saw let sleeping dogs lie surely deals with the consequences of violating a pet's slumber. The concept of ***take*** in the **Endangered Species Act (ESA)** addresses the issue too, with concern for preventing disturbance rather than protecting privacy. The effect is the same, but there are fundamental differences in assumptions and expectations regarding take and privacy.

British academic Brett Mills wrote: "We can never really know if animals are giving consent, but they do often engage in forms of behaviour which suggest they'd rather not encounter humans."[14] Animals don't seem to object to being watched until someone approaches too closely or violates their **flight distance** (Figure 11.1). Their assertive or evasive actions then tell us their privacy has been infringed. Modesty, the human attribute ensuring privacy, is not an attribute of animals. My dog has no sense of modesty when he takes his daily constitutional, and I don't think my praise has anything to do with it.

When the popular press characterizes animals as thinking like people, it inadvertently endorses the notion that animals have a right to privacy (see "Self-Awareness, Cognition, and Privacy"). A recent article, "No Pictures, Please! Andean Bears Wreck Research Camera," is a good example.[15] The writer mixed tongue-in-cheek anthropomorphism with biological fact and succeeded in advancing the conservation benefits of camera trapping. I don't take issue with it, but gullible readers will believe that bears object to having their pictures taken.

Privacy has moralistic implications of propriety and decency that don't apply to animals. Should we avert our eyes and quietly walk away when we witness animal courtship? Are the copulating rhinos in the zoo embarrassed that we are watching? Respect for animal privacy would turn us away blushing, but facts of life are matters of daily concern for biologists.

I believe it is easier to evaluate the behavioral consequences of our actions on wildlife than to pass moral judgment. Some of the answers can be found in trail camera images, but we have to examine the camera trapper's bag of tricks to answer questions about risks. The use of scent, bait, decoys, and photographing dens can have consequences that are not captured in pixels. We also have to consider the electronic flash, which has an effect so different from that of objects that animals normally encounter in nature.

SELF-AWARENESS, COGNITION, AND PRIVACY

The concept of privacy also assumes a sense of self-awareness. Mirror self-recognition is the chosen tool of investigating this in wild animals. Basically, scientists try to discern signs of recognition in the animal's reaction to its image in the mirror (Figure 11.2). Most species that have been tested give no hint of self-recognition. A chipmunk may stop gathering sunflower seeds to peer briefly at its own image, but it soon resumes feeding. A dog may react as it would to another dog, but soon loses interest. Chimpanzees and certain whales, on the other hand, interact with their images in ways that strongly suggest they recognize themselves. Subtle experimental changes in the subject's appearance, for example, elicit telling responses. These reactions may be manifestations of self-awareness, but they don't reveal whether mammals and birds have a sense of privacy. Privacy, therefore, is a questionable cognitive ability in most mammals and birds; and if it exists there, it is unlikely to have the emotional nuance known to humans. It is easier to determine if our activities endanger animals than invade their privacy.

Figure 11.2. A mirror set can elicit surprising and puzzling results. This cat bolted a moment later, as though it saw a strange interloper. (Butte County, California; video screenshot, GoPro Hero2)

AFFECTING ANIMAL BEHAVIOR

"Among the principal dos and don'ts of responsible nature photography is the requirement that the photographer not affect the behavior of the subject in any way," wildlife photographer Gustav Verderber wrote in an article about camera trapping.[16] An animal that detects the photographer may flee or suspiciously fixate on the photographer. Neither situation yields an interesting variety of images.

It is practically impossible, however, to set a trail camera that mammals can't detect at close range. Mammals and birds have adaptive responses to novel stimuli like trail cameras; they investigate them or avoid them and then get on with their lives. There is no scientific evidence that camera traps divert their attention from issues of survival.

The idea that camera traps exasperate, enrage, or cause resentment among animals is not borne out by observation of photos and footage. Curiosity is the most common reaction to remote cameras, and its manifestation in scent-marking, licking, biting, or dismantling depends on the animal. (Chapter 15 discusses these manifestations in black bears.)

Remote cameras do not magically appear in nature. We hike into the boondocks to make our sets, believing that our environmental impact is negligible. Finding that animals visited the set, sometimes within hours of our arrival or departure, reassures us that our footprint is small compared to that of mountain bikers and off-road-vehicle enthusiasts.

The cumulative effects of nonconsumptive human recreation are often subtle, but recent studies show that even nonmotorized human presence can bring about a decline in the density of native carnivores and cause a shift in species composition from native to introduced species.[17] Studies of bald eagles have shown that human recreation affects eagle numbers, distribution, activity, and feeding.[18] Ironically, camera trappers are nature lovers who are lumped with hikers, orienteers, and cross-country skiers as just another segment of humanity that is loving nature to death.

As the population grows, more people compete for access to wild landscapes, and they affect wildlife in many ways. More regulation is inevitable, but we can self-regulate our activities according to Verderber's dictum, which I would modify slightly: Responsible camera trapping requires that the photographer not affect the behavior of the subject in any way that places it at risk. Let us now consider the risks.

THE RISKS OF DISTURBANCE

Vertebrates react to people and predators in similar ways. They often pay attention without stopping what they are doing, but they may take flight, give alarm calls, or mob the predator if the presence is perceived as a threat. Reacting to predators diverts time and energy from essential activities, such as feeding, parental care, or courtship and nest-building. The risk-disturbance hypothesis predicts that animals adjust their reactions to predators and people similarly and that their course of action depends on the proximity of the disturbance. Their behavioral decisions are a tradeoff between minimizing the risk and maximizing time devoted to other essential activities. The presence of a camera trap is usually perceived as a nonthreat or a low-level disturbance, but the activities associated with making the set can be far more disturbing.

Figure 11.3*A–B*. Gray foxes commonly squint or shut their eyes in reaction to the low intensity pre-flash bursts produced by homebrewed camera traps set in the red-eye mode. (Butte County, California)

Figure 11.3C. When the camera is not set for red-eye, gray foxes are photographed more often with eyes wide open. (Chimineas Ranch, San Luis Obispo County, California; Sony s600)

Figure 11.3. ***D–E***, Bobcats are similar to gray foxes in their reaction to flash, but they often freeze, as if studying the situation. (Chimineas Ranch, San Luis Obispo County, California; Sony s600) ***F***, This unusually bold bobcat examined the camera wide-eyed and up close, and the camera recorded the retinal blood vessels. (Chimineas Ranch, San Luis Obispo County, California; Sony s600)

THE RISKS OF FLASHES, FLOODLIGHTS, AND INFRARED ARRAYS

Camera trappers disagree about the effects of electronic lighting on wildlife, but nothing explains animal reactions to artificial light better than video footage and photos on the internet. Animals may react to a trail camera's sight, smell, and sound. So it isn't easy to pin an animal's reaction to any particular stimulus without experimentation. However, when video cameras monitor trail cameras, they clearly reveal reactions to the flash. Then we see puzzled animals and animals that could care less. Shy animals retreat, while the curious make a beeline to the camera to investigate.

The white floodlights used for video evince similar responses. Wildlife photographers like Dereck and Beverly Joubert and Alan and Jane Root have captured remarkable footage of natural behavior in Africa using these bright lights.[19] It is hard to conclude that such lighting invariably

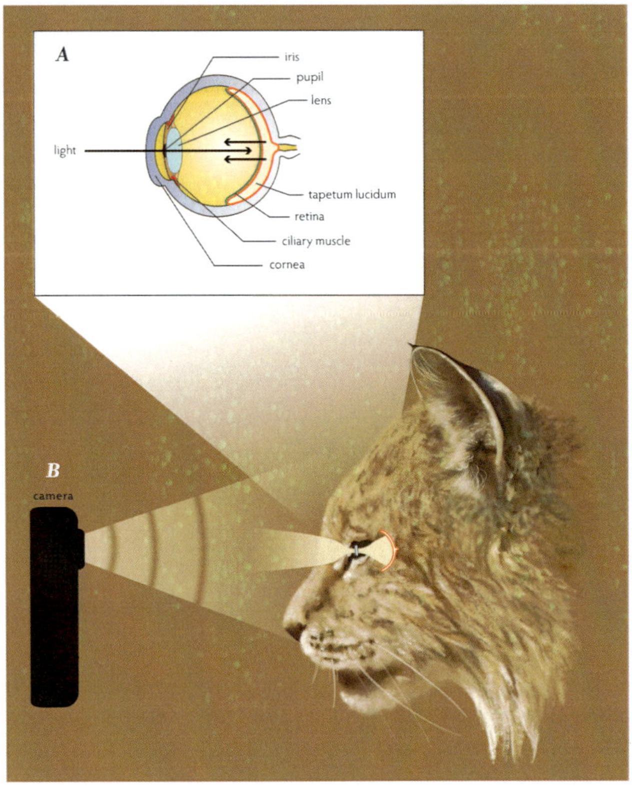

Figure 11.4. A, Camera trapping raises questions about eye anatomy and species differences in eyeshine and vision. ***B***, The reflective area behind the retina known as the tapetum lucidum enhances night vision and causes eyeshine in many vertebrates. *Courtesy of Carol Underwood*

repels or disturbs wildlife. Experience and learning obviously shape an animal's response to artificial light (see chapter 13).

Animals often seem to notice an electronic flash or infrared array at a considerable distance. They may depart or approach in response, but they often delay approaching the camera when investigating scent lures or eating. In general, the electronic flash doesn't frighten most animals. Coyotes and wolves are exceptions. They sometimes take immediate flight and rarely hang around after the first picture is taken. But individuals differ, and young animals are less timid. Wildlife biologist Per Wegge noticed that tigers avoided flash cameras that were triggered by a pressure pad. Apparently, the pressure pad served as a cue.[20]

We are all familiar with the close-range effects of white flash on our own vision. The disturbing but brief impairment to one's sight is due to bleaching of a visual enzyme called **rhodopsin** (see "The Visual Effects of Electronic Flash"). Animals that get a full-face exposure to white flash probably experience the same thing, because they often squint, shut, or avert their eyes (Figure 11.3). However, many species, like bears, rabbits, rodents, and deer, rarely blink an eye.

The red-eye effect in people and the eye shine of many mammals is a consequence of the flash on dilated pupils in dim light. Red-eye is caused by light reflected from our eye's blood-rich choroid layer, while eye shine results from light reflected from tissue in the retina called the **tapetum lucidum** (Figure 11.4).[21] Camera engineers probably developed red-eye reduction to improve photos rather than to spare people the visual artifacts emblazoned on their retinas. In red-eye setting, the camera's low-intensity flashes cause the pupils to contract and admit less light to the retinas (Figure 11.5). Red-eye is also less likely when the gaze is tangential to the camera (Figure 11.6). The result is human eyes that reflect relatively little red light and eyes of mammals and birds that look a lot less like headlights.

Camera manufacturers assure us that a flash in the face is not harmful to people, but what about animals? Flash-blind scavengers at a carcass could be more vulnerable to the attack of larger competitors. Positioning an external flash to the side of the camera usually eliminates eye shine, because the animal is less likely to look directly at the light. Side lighting probably minimizes and may eliminate temporary impairment of vision.

Figure 11.5. ***A***, Eyeshine caused by white electronic flash is subdued in daylight because the pupils are not widely dilated. (Chimineas Ranch, San Luis Obispo County, California; Sony s600) ***B***, This is not the case in deer, whose pupils seem to have a larger diurnal aperture. (Marin County, California; Sony s600)

Figure 11.5. ***C***, Infrared lighting affects pupil dilation negligibly, and "high-beam" light reflection can be expected in night videos. (Sierra County, California; video screenshot, Browning Recon Force Platinum) ***D***, Pre-flash bursts at night cause pupil contraction; reflected light is reduced and the iris is visible. (Chimineas Ranch, San Luis Obispo County, California; Sony s600) ***E***, Absent the pre-flash bursts, this coyote's pupils were wide open and looked like headlights on high beam. (Chimineas Ranch, San Luis Obispo County, California; Sony s600)

Infrared light arrays produce light of lower intensity than white flash, and most mammals and owls show variable responses to the red glow of infrared LED arrays. When viewed from an angle, lighted diodes with wavelengths of 850–900 µm are not visible to humans but may be seen by other mammals and nocturnal birds. The low intensity of this wavelength is unlikely to stimulate a rhodopsin response. Infrared arrays with a wavelength of 960 µm are extremely dim, and even at close range people can barely see the dull red or purple points of light. However, researchers in New Zealand have demonstrated that at least some captive polecats are able to see infrared light in the invisible 960 µm range.[22]

Artificial photographic lighting affects the behavior of different species in various ways, but it is not a strong repellent. We know from personal experience that white electronic flash seen at close range can cause rhodopsin bleaching and a brief disturbing impairment of vision. Animals apparently learn to avoid this reaction. When using homebrewed cameras, we should try to reduce this effect by using the flash's red-eye setting because it reduces temporary visual impairment and any associated disadvantage. Placing external flash units away from the camera is less likely to attract the animal's gaze. Combining these approaches would seem optimal.

THE VISUAL EFFECTS OF ELECTRONIC FLASH

The eye's pupils and a light-sensitive compound called rhodopsin are two features that enable night vision in mammals. Night vision lacks color and is said to be monochromatic. In darkness the pupils dilate to admit light, and the rods of the eye's retina produce rhodopsin. Rhodopsin is highly sensitive to light and changes chemically when eyes adapted to darkness are exposed to bright light. When the night-adapted eye is exposed to light or a white flash, the eyes adjust for day vision. The pupils contract like the aperture of a lens, and rhodopsin undergoes an almost instantaneous chemical change that extinguishes the ability to see in the dark. Humans see a bleached image that gradually fades away. The temporary blindness caused by the flash disappears as the eye again adapts to darkness by producing rhodopsin. The effect is certainly disturbing to people looking directly at the light source, especially at close range, but it is short-lived. Presumably, it has a similar effect on mammals and birds.

USE OF BAIT, SCENT, AND SOUND

Using bait always has ethical implications for hunting, wildlife observation, and photography, but the temptation to use it is often compelling. British wildlife photographer Mark Carwardine wrote: "Almost everyone has baited wildlife at one time or another—even if it's merely putting out food on a bird table. But doing so for photography comes with great responsibility."[23]

The use of unnatural foods as baits can lead to acquired tastes and harmful consequences; nuisance bears living near people are a good example. They commonly encounter pet food and human garbage and often break into houses to find it. Homeowners who use bear-proofing garbage tactics during the week find themselves outsmarted when the bears visit the unprotected cans on trash collection day. Infrequent and limited amounts of unnatural foods do not create a serious problem. Throwing breadcrumbs to a chipmunk or jay will not endanger it, but animals may tempt fate when they steal the lunchmeat or chew their way into a backpack for a granola bar.

Prolonged baiting with natural foods can also disrupt normal foraging patterns and create a dependency, but effects differ between species. Rodents and **corvids** are food hoarders that resort to their larders when a long-standing food supply disappears. Ungulates don't have this option and must search for less common foods. Responsible camera trappers can make use of natural windfalls of food and natural baits without tempting the fate of bears and endangering human property. Obviously, natural foods in an animal's diet are preferable baits to manufactured or human foods.

Figure 11.6. A slight difference in the angle of light rendered this great horned owl's right pupil red and the left one black. (Chimineas Ranch, San Luis Obispo County, California; Sony s600).

DENNING WOLVES AND SLEEPING BEARS

Studies of wolves' reactions to human disturbance in Europe, India, and North America reveal quite a few similarities. First, wolves that den near human habitation are less disturbed by human activities than those in remote areas. Second, parents do not abandon offspring when disturbed, but they may move them to a different den. Third, mild and brief disturbance, such as people walking around the den, is less likely to cause den relocation than capturing and handling pups. Fourth, relocation of a litter to another den depends on the age of the pups at the time of disturbance. Mothers are less likely to move pups who are younger than one month old. Fifth, relocation of the pups occurs within twenty-four hours of disturbance. Sixth, the distance moved did not differ between disturbed and undisturbed litters. And seventh, disturbance doesn't affect use of a den site in subsequent years.

Black bears hibernate in dens for several months in northern latitudes. They give birth, nurse young, and have limited energy stores to compensate for energetic costs caused by disturbance. Researchers have found several predictable patterns. First, bears that den within a mile of human activities are usually unresponsive to disturbances up to a half mile away. Second, responses to disturbance are geared to the costs of moving to another den, which are great in severe climates and minor in mild climates. Third, tree-denning bears tolerate human approach more often than ground-denning bears. Fourth, bears change dens due to natural causes, but disturbances early in the season are more likely to cause abandonment. Fifth, bears differ individually in tolerance to disturbance by researchers. Some submit to the handling of cubs, while others move to another den and even abandon their young. And sixth, bears that abandon dens experience greater weight loss than those that stay put.

Baiting with live animals is distasteful, and most people consider it unethical. Wildlife biologists weigh the pros and cons before using live baits, and in some studies, the ends justify the means. Live prey surveys, for example, work well to locate nesting spotted owls and fledgling owlets that don't normally respond to dead baits, and the owls quickly dispatch the mice.[24] The use of live bait is controlled by state game regulations and is a common practice with falconers, but even where it is legal, recreational camera trappers should avoid the practice.

Scent, visual, and sound lures will not reward you with as many photos as baits will, but unlike baits they are not habit-forming, and until animals habituate to them, their effects on animal behavior are modest. I will revisit this topic again in chapter 13 when examining learning patterns in response to attractants and behavioral deterrents.

NESTS AND DENS

Finding a natal den or nesting birds is a rare opportunity to observe and photograph secrets of life history, but the rearing of offspring is the most vulnerable phase of the annual cycle. There are risks of disturbance, and your activities can have a regrettable outcome if you are ignorant of the species' biology (see "Denning Wolves and Sleeping Bears").

Species and individuals react to disturbance in different ways, and responses change as the young develop. Birds and mammals are more likely to abandon a disturbed nest or natal den before incubation or birth; but once they have offspring, their only option is to protect their genetic investment. The course of action depends on the threat. Each has its risks.

Disturbance can increase the chance of attracting predators. Wildlife biologists discovered many years ago that discreetly counting eggs in waterfowl nests ushered predation by raccoons, foxes, coyotes, and skunks. Just breaking trails to the nests disturbed vegetation and created a superhighway of scent cues used by predators.[25]

Disturbance can also disrupt normal feeding schedules and expose young to heat or cold. When you approach a bird's nest, you trigger antipredator behavior that has evolved to confuse or distract predators. Birds that have altricial young mob ground and aerial predators, while birds with **precocial** young use distraction displays against ground predators. Nesting plovers feigning injury is a familiar example.[26] The bird's commotion can also attract crows, ravens, and birds of prey, and even if it doesn't, you could step on the camouflaged eggs.

When people and predators visit a natal den, carnivores remain hidden underground or watch nervously at a distance. Denning parents bolt if the predator tries to enter and may put up a bold defense. The terrific strength of maternal instinct is graphically revealed in Judy Lemberg's YouTube video of a red fox defending its den against a badger.[27] It is plain to see that it was highly stressful to the mother.

Mammals do not normally abandon litters that are in good health, but they commonly move the young one by one when disturbed—sometimes for no apparent reason. No coyote will forego the chance to chase a mother fox carrying young or to dig up untended young in a den. Nature is red in tooth and claw and doesn't need the help of a bumbling camera trapper to make it any bloodier.

How do you camera trap a natal den without creating a disturbance? Let's assume that you have found a den, have observed a coyote or badger several times in the vicinity, and have finally seen it digging in the entrance from afar. First, do your homework. Read about the season of birth and development of young. If possible, consult a wildlife biologist or graduate student who is studying the species in your area. Second, use a camera with adequate backup power and a large memory card, and set it before the breeding season to allow occupants to adjust to it. Third, visit briefly and move cautiously when the animals are least active—usually midday. Try to visit before and after the most vulnerable phases of den life—those weeks when the young are mobile but not yet using multiple dens. Don't return until the young begin to move between dens. If you are camera trapping animals habituated to human activity, their behavior may be exceptions to the rule (see "Exceptions to the Rule").

Figure 11.7. ***A***, When habituated to cameras and human scent, red foxes offer rare glimpses of parental care at their dens. ***B***, Personality differences become evident when pups are in their natal coat. ***C*** and ***D***, At the age of six weeks, pups have molted and play occupies much of their time. ***E***, Sex differences in behavior also emerge. ***F***, The line between aggression and play becomes blurred as they mature. (Lincoln County, Minnesota; Sony s600)
Courtesy of Chuck Gackstetter

EXCEPTIONS TO THE RULE

Animals habituated to people are much more forgiving of human disturbance, and others are naturally tolerant. Biologist Gary Fellers discovered a long-tailed weasel den while surveying mountain beaver burrows in California. The high jinks of the youngsters indicated they were oblivious to the camera. Chuck Gackstetter photographed denning red foxes in southwest Minnesota for several years (Figure 11.7). These farmland foxes are habituated to tractors, people, and the camera staked next to the den. The first year, the female moved the pups to different dens nearby, but then tolerated the cameras he set at each den.

Figure 11.8. Hyperactive coyote pups exhausted this camera's batteries in twenty-one hours. (San Mateo County, California; Sony s600) *Courtesy of Ken Hickman*

Ken Hickman set a camera a several feet from a coyote den in a culvert when the pups were about a month old (Figure 11.8). He returned three weeks later. He wrote:

> *Of course, it didn't go exactly as I expected, i.e., seeing them get bigger, ma and dad bringing food, etc. Instead, the seven (yes, seven!) pups came out to play so much, that the cam took 917 photos and shot the batteries out in twenty-one hours—from 7:00 p.m. one day 'til 4:00 p.m. the next. So, I only really caught a snapshot in time. Well, 917 of them . . . Trap and learn.*

Take-home message? Prepare for a plethora of pictures, use the largest memory card possible, and have high-capacity backup power. You may still have to service the camera before the litter disperses.

TRUTH AND DISCLOSURE

"One must be willing to declare the process of making their images; it is an act of essential self-awareness," wrote *National Geographic* photographer Michael "Nick" Nichols. "I firmly believe that not revealing the process leads to darkness whether or not the truth is eventually exposed."[28]

That darkness settled on a Spanish wildlife photographer in 2009 when his prize-winning entry of a wolf breaching a rustic gate was deemed a fake. Biologists doubted that a wild wolf would jump a gate if it could squeeze through with less effort, and critics identified the animal as a "wildlife model." The BBC's Wildlife Photographer of the Year judges changed their minds and rescinded the prize.[29]

It's hard to fool wildlife biologists with faked wildlife photos, but the public is more easily duped. Nature-faking has been a fact of the wildlife film industry at least since the early days of Disney's wildlife productions. Popular filmmakers passed off myth as reality. Tricks of the trade included fabricated natural history sets and tame animals rented as animal actors. There is nothing inherently wrong with artificial film sets or using healthy captive stand-ins to create near-impossible footage, but in those days, filmmakers kept the techniques secret. Ironically, wildlife conservation organizations promoted their cause with stunning photos of wildlife, some taken in zoos and game farms.[30]

Exposure of nature fakery in several famous films turned the tide of public opinion, and the media started to discreetly disclose the unnatural conditions. Disclosure is now the norm in mainstream media, and as writer Ted Williams notes, "These days there is little that I (or anyone) can positively identify as nature fakery or animal abuse."[31]

Trail camera hobbyists are not subject to production deadlines and high-stakes temptations. They satisfy their ego needs by sharing photos and information on the internet. Information is freely shared, and newcomers are encouraged in trail camera forums. When contentious issues surface, the threads may be entertaining, but ethics prevail. When someone asks what bait or scent was used, the answer is forthcoming. Post a questionable photo, and someone will call your bluff.

There are several reasons for not disclosing photographic techniques or manipulations. First, the photographer might have forgotten or thought the explanation was obvious. Photos of scent-marking mammals are almost always responses to scent lures, and this is often not disclosed. In such cases, disclosure of natural scent-marking seems to make more sense. Second, we can't expect anonymous pranksters to disclose anything about the unlikely images they Photoshop for the gullible. Sooner or later someone denounces a faked photo for what it is, but fakes have a way of re-emerging when a new audience discovers them. Unfortunately, we never find out if some improbable images were faked. Third, a photographer won't disclose illegal methods or announce manipulation. And finally, a good liar has no qualms about disclosing a creative fabrication.

Truthful disclosure builds trust, educates, and sets an example of professionalism. The many high-quality images taken by amateurs prove that cooperation and integrity pay off. The benefit of cooperation and integrity among camera trappers is the ever-growing number of high-quality images on the internet.

FAIR CHASE

Fair chase, or ethical hunting, means hunters pit their knowledge, skills, and endurance against the wit and instinct of the quarry, within the framework of the law. The philosopher José Ortega y Gasset wrote: "The inequality between hunter and the hunted should not be allowed to become excessive."[32] Communion with nature is as essential to the process as slaying the quarry.

Technology doesn't necessarily keep the chase fair, but when high-power rifles with optical gun sights are available, few hunters opt for handmade weapons. We tend to embrace new hunting technology without meditating on ethics. When hunting guides offer drones to locate prize trophies, big-game hunters will seize the opportunity—and Teddy Roosevelt will turn in his grave. That great advocate of "the strenuous life" would most certainly rail against drones as giving lazy hunters an advantage they haven't earned.

Few state game agencies have found it necessary to regulate the use of trail cameras for hunting. The remote camera aids the hunter in gathering intelligence, and the agencies have deemed that advantage acceptable—perhaps to lend the sport wider appeal, perhaps to give unskilled hunters a better chance to bag an animal. That is beginning to change. In 2010 Montana prohibited using trail cameras during hunting season (see chapter 12).[33] The law was enacted for the sake of fair chase. Cellular trail cameras, which can email photos to the hunter, were of particular concern. Three more states—Arizona, Nevada, and Utah—have followed suit to curb widespread abuse of fair chase because hunting and outfitting businesses have swamped water holes with trail cameras and marketed photos for commercial gain. The unfortunate consequence of these laws is the loss of recreational and educational opportunities for nonhunting camera trappers for several months, including winter, the best time to photograph many predators.

No one has data showing how trail cameras affect hunter success, but any university wildlife department, nongovernment organization, state game department, or cooperative fish and wildlife research unit could clarify the issue by doing the research. Until then, with or without trail cameras, hunters have to determine their own ethical boundaries within the legal framework. As Aldo Leopold noted: "A peculiar virtue in wildlife ethics is that the hunter ordinarily has no gallery to applaud or disapprove of his conduct. Whatever his acts, they are dictated by his own conscience, rather than by a mob of onlookers. It is difficult to exaggerate the importance of this fact."[34]

CHAPTER 12

Legal Issues

> Ignorance of property and game laws can turn camera trapping into a bad dream. Setting cameras on private, state, or federal lands requires permission, and some camera trapping activities may violate state and federal laws regarding game and endangered species. Game laws differ between states, but every camera trapper should be aware of their state's regulations and should confirm them when crossing state lines. The concepts of intent, take, and harassment are relevant legal issues, and so is the use of various attractants. This chapter reviews several sections of California's wildlife code to explore implications for camera trapping.

Most people view camera trapping as a benign pursuit. Whenever news media report new camera trap discoveries, they reinforce favorable impressions. In the age of mass communication and social media, however, reports of misguided camera trapping can easily tarnish the positive perception of an otherwise educational and gratifying outdoor recreation.

Wildlife photographers must know wildlife laws. Consider our need to find undisturbed land to photograph wildlife. An ideal location may be privately owned or controlled by public agencies and subject to the laws of trespass. Suppose we are camera trapping an endangered species. Our activities are subject to international, federal, and state statutes, and we should understand the concept of *take*. Finally, the overzealous use of bait is a concern of state game wardens because it can disrupt the foraging behavior of wildlife. Camera trappers who comply with state and federal wildlife law and respect property rights do not have to invoke ignorance of the law as an excuse. It doesn't wash with law enforcement officers anyway.

This chapter discusses the legal authorities that regulate the use of wildlife, wildlife and game law, and land ownership. It isn't practical to summarize legal issues in every state, but here we examine the implications of California's state game code to camera trapping as an example of legal issues that may apply in other states.

Figure 12.1. Federal and state agencies post the land in their jurisdictions, and permission to camera trap may be necessary.

AUTHORITIES CONTROLLING PUBLIC LAND AND WILDLIFE

There's a good reason why fish and wildlife in the United States are held in the public trust. For several centuries our forefathers believed that natural resources and wildlife were without limits. Unbridled market hunting and extermination of the passenger pigeon and other species proved otherwise, and the government deemed wildlife and natural resources "too valuable for the greater good to be held in private ownership." Over the next hundred years, numerous pieces of legislation for the conservation of land and wildlife came into effect.

The use and management of fish and wildlife in the United States are controlled by the federal government and individual states. Two bureaus of the US Department of the Interior regulate fish and wildlife. The US Fish and Wildlife Service (USFWS) is the regulatory agency for the management of fish, wildlife, and natural habitats, and the US National Park Service (NPS) and Bureau of Land Management (BLM) manage protected areas and rangeland (Figure 12.1).

The US Forest Service (USFS) is a branch of the US Department of Agriculture (USDA) with oversight of almost two hundred million acres of national forests and grasslands. It protects and manages the nation's forests, rangeland, and aquatic ecosystems for recreation and extractive use. It also partners with private landowners and state agencies to provide technical and financial assistance. Livestock grazing, hunting, timber felling, and mining are common legal activities on USFS and BLM lands.

In addition to federal land management agencies, state-owned lands are managed by state fish and game agencies, state forests and parks, and county and municipal parks. Indian Tribal governments administer Native American lands.

PROPERTY RIGHTS

Never assume that a quiet, forgotten piece of rural America belongs to no one. All of it belongs to someone. It is either private or public land, and you will commit criminal trespass if you knowingly enter private property without consent or written permission from the owner or the person leasing the land. Owners of rural property usually post their land with signs and fences, and some landowners are receptive to the idea of hosting a camera trap survey. Country folk know their neighbors and are good sources of information about local property, but the county records office is also a good source of information on land ownership.

Trespassing is a misdemeanor in most states, but landowners must rely on law enforcement officers to apprehend trespassers. Landowners are within their rights to appropriate a camera trap found on their property (Figure 12.2). In the state of Texas, on the other hand, landowners can lawfully fire upon trespassers at night.[1] For camera trappers who hunt, many fish and game codes require written permission to be carried with them while hunting. Obviously, it is good practice to get written permission to camera trap on private property.

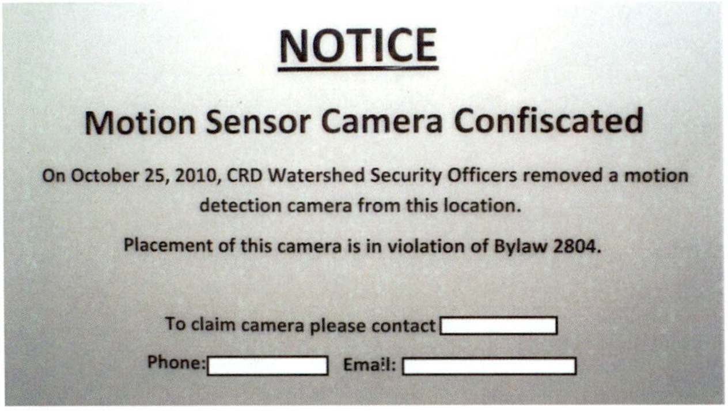

Figure 12.2. Camera trapping without permission may result in camera confiscation and a citation.
Courtesy of Gary Schroyen

You need permission to set cameras on federal and state wildlife refuges, state and county parks, and national parks. Getting permission may be as simple as sending an email request, or paying a visit to the refuge office, but usually you will be asked to make your request in writing, or via an online permit form. Be prepared to wait a few months before getting approval and guidelines. Permission may be needed from several agencies if a reserve has multiple jurisdictions. Permits for access to conduct fieldwork such as camera trapping are often called Scientific Collection Permits. Agencies often require the camera trapper to sign a liability waiver to protect themselves from accidents during fieldwork. Corporate landowners such as timber companies may want proof of insurance coverage for personal liability, and the minimal amount of coverage may be more than your policy allows. However, they may allow you to camera trap if you sign a disclaimer.

Permission is not required to use camera traps on BLM or USFS lands,[2] but permits are required on those designated as "Research Natural Areas" and "Areas of Critical Environmental Concern." Since laws change, it is a good idea to consult with the regional forest service office or ranger district headquarters for approval and advice. The staff biologist may be interested in your findings, and they can inform you if controlled burns or timber extraction is scheduled in areas of interest. The National Park Service, on the other hand, requires permits to use remote cameras for scientific research, and you will be required to write a proposal to explain the purpose, methods, and benefits to the agency.[3] Permission is also necessary to set cameras on national wildlife refuges. Since the park service often conducts wildlife surveys with trail cameras, they can certainly be approached to explore the possibility of participation.

FEDERAL WILDLIFE LAW

Laws regulating the protection and use of wildlife are encoded in federal[4] and state legislation, namely the ESA and state fish and game codes, and apply to all wildlife, including on private lands. In 1973 the US Congress passed the ESA to save animal and plant species from extinction, preserve ecosystems critical for their survival, and provide programs for their conservation. The National Marine Fisheries Service and the USFWS are the federal agencies that administer the ESA for marine and terrestrial species, respectively.

ON DISTURBING ENDANGERED SPECIES

The San Joaquin kit fox is an endangered species that inhabits a region rich in oil development (Figure 12.3). The US Fish and Wildlife Service (USFWS) and state of California regulate all development activities in the fox's habitat under the Endangered Species Act (ESA) and the California Endangered Species Act (CESA).

Because the kit fox habitat is vital for the species' survival, authorized development must follow measures designed to protect foxes before and during ground disturbance and construction activities. These regulations give an appreciation of the concerns and the law.

Only qualified biologists are authorized to survey and monitor populations and preside over destruction of dens. Qualified biologists are persons "who have completed at least four years of university training in wildlife biology or a related science and/or have demonstrated field experience in the identification and life history of the San Joaquin kit fox" and other wild canids.

Figure 12.3. A den of the endangered San Joaquin kit fox at the Lokern Ecological Reserve, Kern County, California. The species benefits from protections under federal and state Endangered Species Acts. *Courtesy of Samantha Bremner Harrison, ESRP (Endangered Species Recovery Program), California State University, Stanislaus*

Preconstruction surveys must be conducted for any project activity likely to affect the San Joaquin kit fox. Habitat should be identified and evaluated for use, and, if possible, potential impacts should be evaluated.

The status of all dens should be determined and mapped, and the biologist must notify the USFWS immediately when a natal den is discovered in a development project or within 200 feet of its boundary.

Whether active or inactive, natal or pupping dens may not be disturbed, and a 100-foot exclusion zone must be marked around known natal dens. Project boundaries should also be marked, or protective exclusion zones (allowing foot traffic but not motorized vehicles) should demarcate all known and potential dens outside the project boundary.

A

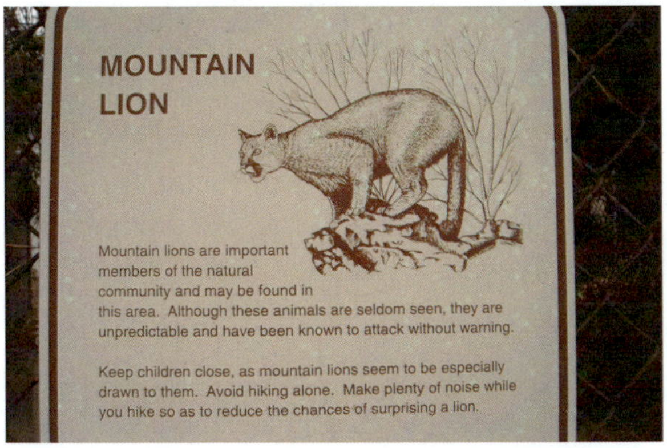

B

C

D

Figure 12.4. In Canada and the United States, signs about wildlife are intended to inform and protect the people. ***A***, Play it safe around plague infected prairie dogs. These signs were posted near Badlands National Park, South Dakota. ***B***, Mountain lions are common in much of California, and so are people. Coexistence means understanding the risks. ***C***, Hikers beware—leave your dogs at home and don't tempt the wolves. ***D***, Tourists enjoy watching Roosevelt elk during the annual rutting season. Wildlife agencies do not assume they understand the danger. This sign was in Humboldt County, California.
A: *Courtesy of Dean Biggins*
B: *Courtesy of Gary Schroyen*

The act mandates evaluation of the population status of species based on the best scientific and commercial data available. When it is justified, a species is listed as *endangered*, which means "in danger of extinction throughout all or a significant portion of its range," or *threatened*, which means "likely to become an endangered species within the foreseeable future throughout all or a significant portion of its range."[5]

The US Department of Interior must produce a recovery plan for endangered species and designate habitat that is critical for species survival. The plan must be based on scientific data, with due consideration given to the economic impact of the designation as endangered or threatened.

Figure 12.5. Signs are also intended to inform people about the danger they pose to wildlife. ***A***, A sign in rural British Columbia elegantly warns that human food is an ever-present threat to wildlife. Its packaging is dangerous if eaten, and human food leads to habituation, food conditioning, and conflict with people. ***B***, Slow down, enjoy the scenery, and give Vancouver Island's bears a chance. ***C***, Enjoy wildlife without disturbing it. Signs at Marin County's Bolinas Lagoon explain wildlife disturbance and how to report it to authorities. ***D***, A sign near the Ishi Wilderness Area in Northern California reminds fishers about the law. Note that the tree has escaped penalty for defacing the sign.
A: *Courtesy of Jennifer Bain* **B**: *Courtesy of Gary Schroyen*

The ESA also introduced the concept of *take*, which prohibits all citizens, including private landowners, from harming endangered and threatened species or significantly modifying **critical habitat**. *Take* is broadly defined as an action that kills, harms, or injures wildlife, but it also includes indirect actions that affect resources essential for survival, such as water, food, and shelter (see "On Disturbing Endangered Species"). Activities that disrupt behavior such as breeding are also considered take. Knowingly "taking" and trafficking endangered species are punishable offenses, but ignorance of the law or accidentally taking endangered species is also punishable.

Wildlife agencies and organizations go to considerable trouble and expense to alert people to the risks of interacting with wildlife (Figure 12.4). They also inform them of threats to wildlife posed by human activities (Figure 12.5).

STATE WILDLIFE LAW

State wildlife agencies and commissions review fish and game codes annually and revise them periodically. Each state's fish and game code contains the statutes or laws passed by the legislature or citizens via ballot initiatives and authorizes the game commission to adopt the regulations and interpret their implementation. Fish and game codes are daunting legal documents, but each state makes brochures available to the public through the agency's offices and website. These brochures address regulations for different classes of game (see "Legal Classification of Game and Nongame in California").

Wildlife laws differ from state to state. States maintain their own lists of threatened and endangered species and may list species as threatened or endangered that are not federally listed.[6] Activities legal in some states, like the use of bait, scent lures, and electronic callers, may be illegal in other states. Hunters and trappers are generally well acquainted with their own state game regulations, but nonhunting camera trappers may know very little about the subject (see "Intent and Criminal Law").

Few game codes mention trail cameras specifically, but undoubtedly this will change as the ranks of amateur photographers grow and management agencies respond to the consequences. Montana's game department added a new provision to the game code in 2010.

> It is unlawful for a person, while hunting, to use any electronic motion-tracking device or mechanism that is designed to track the motion of a game animal and relay information on the animal's movement to the hunter. Motion tracking devices are defined by F & W Commission as remote operated camera or video devices capable of transmitting real time information, pictures or videos; seismic devices; thermal imaging devices; and satellite and radio telemetry devices.[7]

Some Montana camera trappers may want to seek clarification: "Officer, I don't hunt, I don't have a hunting license, and I don't allow hunters to hunt on my land. Can I leave a trail camera on my land through the winter? It's not a Wi-Fi camera." It never hurts to ask.

Let us now examine California's fish and wildlife code and regulations. Even if you live in another state, California's code introduces issues pertinent to users of trail cameras.

RELEVANT DEFINITIONS

It's important to understand the relevant definitions. Section 86 of the California Fish and Game Code, for example, defines *take* more specifically than the ESA does.

> "Take" means hunt, pursue, catch, capture, or kill, or attempt to hunt, pursue, catch, capture, or kill.[8]

In Section 25.1 the game regulations also define harassment as follows:

> . . . No person shall harass, herd or drive any game or nongame bird or mammal or furbearing mammal. For the purposes of this section, harass is defined as an intentional act which disrupts an animal's normal behavior patterns, which includes, but is not limited to breeding, feeding or sheltering.[9]

Knowing these two definitions should give some comfort, because a trail camera in the woods clearly cannot take or harass wildlife. Be aware, however, that interpretation of the law is a subjective matter.

BAITING OF WILDLIFE

California prohibits the take of wildlife over a baited area, but the regulations are more complicated. Section 257.5 of the California Code of Regulations defines a baited area as "any area where shelled, shucked or unshucked corn, wheat or other grains, salt or other feed whatsoever capable of luring, attracting, or enticing such birds or mammals is directly or indirectly placed . . . ,"[11] or any attractant "composed of mammal, bird, or fish flesh, fur, hide, entrails, or feathers."[12]

In Section 251.3, the state prohibits feeding of big game animals: "No person shall knowingly feed big game mammals. . ."[13] This regulation is not specific to hunting. Camera trappers in California are prohibited from using bait to attract big game for photography. For example, the use of automated feeders for ungulates is not legal in California (see "Intent and Criminal Law").

A logical question then is: May I use road-killed animals as bait for camera trapping? Section 2000 of the California Fish and Game Code provides the answer:

> Possession of a bird, mammal, fish, or reptile or parts thereof in or on the fields, forests, or waters of this state, or while returning therefrom with fishing or hunting equipment is prima facie evidence the possessor took the bird, mammal, fish or reptile or parts thereof.[14]

In other words, you cannot possess any game animal, furbearer, nongame mammal, or protected species unless you have a hunting license, trapping license, or scientific collecting permit and any required tags or stamps. On the other hand, Section 473 of that same code says you may possess nongame animals such as mice and rats that are legally taken with firearms or snap traps.[15]

The law informs us when the use of bait is legal. We are not prohibited from setting cameras at naturally occurring foods or at salt licks for cattle. It is also legal to set cameras at gut piles of killed game animals, at carcasses found in the woods, or at natural mineral licks or windfalls of wild or cultivated fruit. Big game might be among the animals attracted, but you need not worry. You did not place the bait and did not break the law by setting your camera there.

Bait is also prohibited for taking resident small game in California (see "Legal Classification of Game and Nongame in California"), but this regulation does not apply to camera trapping because it is not considered *take*. In other words, it is not illegal to photograph squirrels or chipmunks feeding on sunflower seeds at your bird feeder, but you can be cited if you do it in a national or state park.

Finally, biologists often use bait as an attractant in camera trap surveys of carnivores because it greatly increases the chances of detecting certain species. For nonbiologists, this is permissible under a scientific collecting permit or a trapping license from California's Department of Fish and Wildlife.

USING SCENT LURES AS ATTRACTANTS

California's Fish and Wildlife Code does not speak specifically about the wide assortment of scent lures commonly used by trappers (see chapter 8). Cocktails of fermented animal tissues and scents, which elicit a feeding response, are therefore defined as baits. Social attractants, which elicit sniffing, sensing with the Jacobson's organ, and scent-marking, are not baits. These include essential oils from plants, such as catnip or **tonquin oil**, and the raw products or extracts from mammalian scent glands. Such trapping products are not always precisely labeled with regard to their contents. However, some of the most effective attractants are simply essential oils and glandular secretions, and it is legal to use such attractants for camera trapping in California.

INTENT AND CRIMINAL LAW

A legal definition of ***intent***: "the exercise of intelligent will, the mind being fully aware of the nature and consequences of the act which is about to be done, and with such knowledge, and with full liberty of action, willing and electing to do it."[10]

The importance of understanding game and fish regulations came to me in a visit from the game warden, who had read my blog and informed me that it was illegal to use bait to attract game animals in the state of California. I had smeared syrup on a tree trunk to induce a bear to climb. Being a law-abiding citizen who didn't know the game laws, I was greatly embarrassed, but I didn't understand how I could be guilty of this misdemeanor when my neighbors routinely attracted bears by leaving dog food on their porches.

"It all boils down to intent," said the warden. "Your neighbors aren't trying to attract wildlife. They're ignorant and don't know better, and they're creating an attractive nuisance. You are actually trying to attract bears. That's your intention, and that's why you are breaking the law and they aren't."

He didn't cite me, but I learned my lesson.

LEGAL CLASSIFICATION OF GAME AND NONGAME IN CALIFORNIA

- **Resident game birds**: doves, quail, grouse, ptarmigan, partridges (introduced species), pheasants, wild turkeys
- **Resident game mammals**: jackrabbits, varying hares, cottontail rabbits, brush rabbits, pygmy rabbits, tree squirrels
- **Big game mammals**: deer, elk, pronghorn antelope, wild pig (feral domestic, European wild pig, and hybrids), black bear, Nelson bighorn sheep
- **Furbearing mammals**: pine marten, fisher, wolverine, mink, badger, river otter, gray fox, silver fox, cross fox, red fox, kit fox, raccoon, beaver, muskrat
- **Nongame animals**: English sparrow, starling, American crow, coyote, weasels, skunks, opossum, moles, rodents (excluding tree squirrels and flying squirrels), fallow deer, sambar, sika and axis deer, aoudad, mouflon, tahr, and feral goats
- **Protected mammals**: Morro Bay kangaroo rat, bighorn sheep (except Nelson bighorn), southern sea otter, northern elephant seal, Guadalupe fur seal, ring-tailed cat, wolverine, salt marsh harvest mouse, Pacific right whale

USING RECORDED SOUND AS AN ATTRACTANT

The use of recorded animal sounds or imitations of such sounds to *attract* wildlife is not prohibited by Section 3012 of California's Fish and Game Code, but only a small number of nongame species can be *taken* using wildlife callers. The code reads:

> It is unlawful to use any recorded or electrically amplified bird or mammal calls or sounds, or recorded or electrically amplified imitations of bird or mammal calls or sounds, to assist in taking any bird or mammal, except nongame birds and nongame mammals as permitted by regulations of the commission.[16]

Section 475 of the California Fish and Game Code has this to say:

> Recorded or electrically amplified bird or mammal calls or sounds or recorded or electrically amplified imitations of bird or mammal calls or sounds may not be used to take any nongame bird or nongame mammal except coyotes, bobcats, American crows and starlings.[17]

This regulation may be interpreted as not applicable to camera trapping because it is not a form of taking game. But a warden might deem a long-playing sound recording as a distraction that alters animal behavior.

USE OF DECOYS AS ATTRACTANTS

The common practice of using decoys to hunt game birds is legal as long as the decoy is not a living creature, so it is legal for camera trapping as well. The code does not discuss other types of decoys, such as rubber snakes and owl models used to frighten birds from fruit orchards. While these might be used legally, and animals are known to quickly habituate to them, there are ethical reasons to be cautious about using them because they can interfere with time budgets, foraging, and care of the young in birds.

Though camera trapping is an activity with a light ecological footprint, and most state game codes do not address camera trapping in detail, trail camera users should know state wildlife regulations pertaining to their activities. The regulations may not answer all of your questions because some issues are subject to interpretation, but contacting the department of game and wildlife should help clarify any concerns or uncertainties

CHAPTER 13

Animal Psychology for Camera Trappers

Knowing how animals learn helps us interpret and predict animal behavior. The fear of unknown things (**neophobia**), and the waning of responses to inconsequential things (**habituation**), are learning abilities shared by many animals. Animals also learn associations between different stimuli and experiences, and rewards and punishments reinforce or strengthen learned responses. Learning also influences whether animals react the same way to similar stimuli (**stimulus generalization**) or discriminate between stimuli. I discuss these and other kinds of learning in relation to camera trapping, predator control, and conflicts between bears and people in national parks.

An agitated colleague notified me once about a disturbing incident of vandalism. She was doing her first ecology project, a capture and release study of rodents, and her field assistant had just reported bad news. Someone had meddled with all of her live traps and released the trapped mice. She suspected the culprit was one of my employees known for practical jokes, and she was going for bear. I squirmed when she told me my responsibility was to prove the case and take disciplinary action against the offender.

"Wait a minute," I stammered. "Tell me what happened again." After taking time to collect my thoughts, I offered an alternative explanation: "It sounds like a **trap-happy** raccoon found your traps." Curious raccoons are the bane of biologists who use live traps to study rodents. When a raccoon makes the connection between a live trap and an easy meal, it searches for other traps with single-minded purpose. "It probably ate some of your study animals, too," I added. She was neither amused nor convinced but allowed the possibility. Other colleagues persuaded her that any personnel action should be directed to the raccoons.

Figure 13.1. This young adult raccoon's reaction to a homebrew camera and white flash changed from caution to confidence. ***A–C***, It triggered the camera fourteen times over twenty minutes from the safety of the burrow. ***D***, Then it emerged, studied the camera, and left the scene. (Butte County, California; Sony s600)

Matching wits with wildlife can be humbling. Wrong assumptions about animal behavior often hinder predictions, as when camera trappers meticulously stage a creative shot and learn later that the subject thwarted the plan. Was it bad luck, was it an accident, or was the animal just smart? Experiences like this teach us that we often underestimate the perceptual and learning abilities of our subjects and that understanding their cognitive abilities is an asset.

This chapter discusses some concepts of learning theory. It's a subject that has long been the realm of comparative psychologists who have studied domestic rodents and birds in the laboratory.[1] Fewer studies have examined learning in wild animals, but intriguing and amusing examples of learning can be inferred from numerous written accounts and anecdotes. The topic may seem of marginal interest to some camera trappers. But if you have ever wondered if your camera trapping subjects are playing with your mind, this chapter may help you appreciate their cognitive abilities.

NEOPHOBIA: FEAR OF THE NEW

My dog was six months old when we chanced upon a road-killed gray fox. It was a beautiful corpse, lacking gore, but Fred immediately balked and pulled away on his leash. His fear gave way to curiosity when he saw me examining the carcass, and after a cautious sniff he decided it was a plaything. His initial response was a clear example of neophobia, an inborn fear of new or unfamiliar things.

Neophobia is distrust of potentially dangerous organisms, objects, foods, or even smells and circumstances. Body language conveys avoidance, apprehension, or ambivalence. Mammals often hesitate, lift a foreleg (the "tripod stance"), focus their senses, or vacillate between cautious approach and retreat. Birds often bob, show intention movements to take flight, or simply startle and fly away. The cause of fear may not be obvious. Naturalist Bernd Heinrich found that ravens show fear when "events violate expectation." When ravens experience a common event out of context, like a jiggling bush on a windless day, they find it to be very scary.

Trail cameras may alarm animals that have never been exposed to them (Figure 13.1).[2] The images show them lurking in the distance or beating a hasty retreat. Reactions may differ between species and between individuals and may also differ with sex, age, and social status. All populations have outliers: individuals that don't fit the norm (see "Coyotes: Bold and Shy"). There are general taxonomic differences, too. Rodents are often bold around trail cameras or may seem oblivious, but coyotes are usually shy of them.

COYOTES: BOLD AND SHY

Texas historian J. Frank Dobie wrote: "Shyness and curiosity duel with each other in many animals." Both traits are adaptive, but animals tend to favor one or the other. At the wrong time and place either tendency can have fatal consequences.

Camera trappers agree that coyotes are shy subjects. They seem suspicious of our cameras, usually keep their distance, and rarely display the curiosity of a bear or raccoon. Eveline Sequin and her coworkers studied wariness in wild coyotes and found they are not all the same (Figure 13.2). Coyotes are keenly aware of our presence and cameras, but their vulnerability to being photo-captured depends on social status and location with respect to territory.

An understanding of coyote society helps put these findings into perspective. Mated pairs of coyotes share their territory with their offspring of one or more generations and defend it against intruding neighbors. Parents are socially dominant, enjoy priority access to resources, and are thus called *alphas*, while their lower-ranking adult offspring, or *betas*, assist in rearing the pups. Between adjacent territories dwell transient coyotes—landless loners that furtively eke out a living in the borderlands of coyote society. Territories have heavily used central areas and less frequently used peripheral areas. Thus, all coyotes can make forays into adjacent territories without necessarily being encountered.

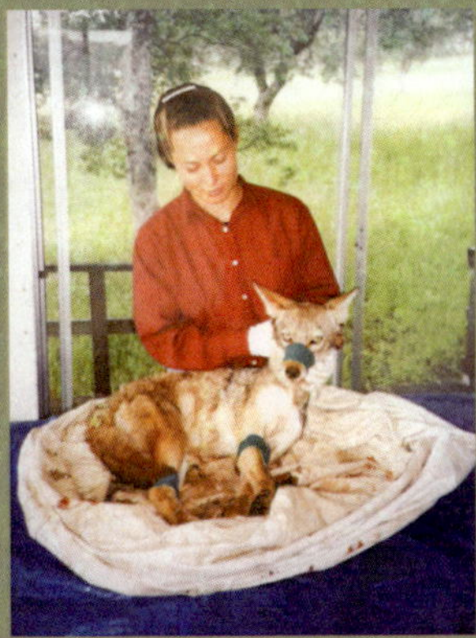

Figure 13.2. In this photo from 2000, Eveline Larrucea (nee Sequin) radio collars a newly captured juvenile during her study of coyote responses to camera traps. During handling, the animals were restrained with vet wrap. *Courtesy of Eveline Larrucea*

Sequin's team captured, aged, marked, and radio-collared coyotes in a wildlife preserve in northern California and studied their reactions to twenty-four-hour camera traps using AIR motion sensors that didn't allow false triggers. They set cameras without bait or scent lures on roads and game trails. They never photographed a coyote at the same camera more than once within the same hour, and 83 percent of their photos were taken between sunset and sunrise. More surprising was the finding that age had no significant effect on photo-capture of one- to six-year-old radio-collared coyotes. Nor did the presence of a researcher's dog influence rates of photographing coyotes afterwards.

Alpha coyotes often kept tabs on the researchers working in their territories. Sometimes they retreated, but often they were vigilant, and from time to time they even approached for a better view. Transients, on the other hand, always disappeared. Cameras didn't affect any coyote's use of space, but alphas apparently skirted cameras wherever they found them and were rarely photographed compared to betas and transients. Beta coyotes were photographed in and out of their territories, and transients were almost always photographed on territory boundaries.

Alphas seem to know their territories well, remember camera locations, and remain wary of them. Their neophobia to cameras didn't wane during twenty-one months of study, but betas and transients were bolder around cameras and were photographed more often than expected. These findings agree with other studies: coyotes are more cautious in familiar areas than when away from home.

A

B

Figure 13.3*A–B*. Images of coyotes in flight are commonly taken where they are hunted or have not habituated to camera traps. These youngsters instantly responded to the camera's low intensity pre-flash bursts.

C

Figure 13.3*C*. Ambivalent body language. The animal was drawn to a punctured can of fish but was leery of the camera.

D

Figure 13.3*D*. The camera startled this adult coyote while it sniffed a burrow; it was well on its way when the next picture shuttered.

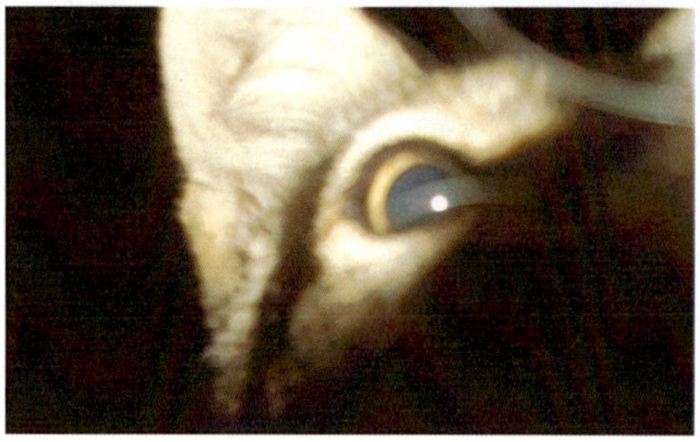

E

Figure 13.3*E*. This bold juvenile showed no alarm to the flashing camera.

F

Figure 13.3*F*. This curious juvenile coyote approached the camera to investigate. (Chimineas Ranch, San Luis Obispo County, California; Sony s600)

HABITUATION: GETTING USED TO HARMLESS THINGS

Many camera trappers witness a common form of learning when they notice that an animal's initial wariness or curiosity about a camera gradually wanes and disappears (Figure 13.3). My first photos of gray fox were few and far between, but as time passed the foxes became bolder, left more photos, and eventually inspected the camera up close. One or two foxes even ventured to paw it. When repeated encounters with strange objects prove harmless, animals lose their fear of them. Animals may also habituate to stimuli that elicit curiosity. In both cases, the stimulus is neutral or inconsequential.

Mammals also habituate to scent lures. My use of castoreum eventually failed to elicit scent rolling in local gray foxes, but when I replaced it with tonquin oil they gave the full-blown response. The mobbing reaction of birds to a stuffed hawk or owl (even a poor plastic model of one) is a thrilling demonstration of antipredator tactics and habituation. I still remember a gathering swarm of alarmed songbirds dive-bombing a stuffed hawk on a field trip in animal behavior. When the commotion petered out, the instructor re-energized the birds a bit longer simply by moving the model to the next fencepost.[3]

It's easy to understand why habituation to harmless new objects is adaptive. Persistent neophobia to inconsequential signs of ATV enthusiasts, hunters, and hikers could greatly hamper wildlife movements. Habituation prevents that from happening.

ASSOCIATIVE LEARNING: LEARNING CONNECTIONS BETWEEN DIFFERENT THINGS

Animals often change their behavior when they notice relationships or associations between stimuli, or between a stimulus and a consequence. One familiar kind of **associative learning** is **classical conditioning**, discovered over a century ago by Russian physiologist Ivan Pavlov. Pavlov noticed that the dogs in his laboratory salivated when they heard keepers preparing their food. His explanation was that they associated the sounds, which he called a conditioned stimulus, with the smell and taste of the food, the unconditioned stimulus, which normally causes salivation. Any arbitrary signal repeated at feeding time, even the ringing of a bell, could be made to trigger the involuntary reflex of salivation. In other words, a conditioned stimulus can acquire significance if associated with another stimulus of consequence.

Classical conditioning deals with involuntary reflexes, but animals also learn to modify voluntary behavior to specific consequences. As anyone who has trained a pet knows, rewards or positive reinforcement can increase a pattern of behavior, while punishment or negative reinforcement can decrease it. This is called **operant conditioning** because learned responses operate upon, or affect, the animal's environment. Pet owners are keenly aware of associative learning. Their pets' attentiveness to their own routines and their uncanny ability to anticipate events are endearing reminders of this ability.

Reinforcement, that is, reward or punishment, affects responses in predictable ways. Animals will work harder for preferred rewards, such as palatable foods or larger servings. Bird lovers who attempt to block the access of tree squirrels to bird feeders soon learn about their amazing problem-solving and gymnastic abilities.[4]

DISCRIMINATION LEARNING

Animals benefit from learning to distinguish between similar stimuli. Predators discriminate the palatability of different prey species, and herbivores discriminate the smell, taste, and nutritional content of forage plants. They learn from experience to recognize unique visual, olfactory, auditory, and tactile properties, which is a cognitive process called **pattern recognition**. An example is the ability of birds to discriminate color patterns of distasteful species like monarch butterflies, which led to the evolution of warning coloration (or **aposematism**).

Discrimination learning allows predators to recognize cryptic species of prey by forming a search image of a few subtle but salient cues. Insect-feeding birds discover hard-to-see moths and

A

B

C

D

Figure 13.4. Skunks are good teachers. They hold their Jolly Rogers high as a warning. Their skunked victims associate the dreadful experience with the black and white coloration. ***A–B***, Striped skunks (*Mephitis mephitis*) rush at their foes and stamp their feet. The fetid blast is the final act. ***C–D***, Western spotted skunks (*Spilogale gracilis*) have distinctive banners and perform the handstand before discharging their scent glands. Many skunked victims associate the dreadful experience with the black and white warning coloration and curtail subsequent encounters.
A–C: Chimineas Ranch, San Luis Obispo County, California; Sony s600;
D: Mono County, California; Sony s600; *Courtesy of Ken Hickman*

caterpillars by trial and error. They may not be able to see the whole organism camouflaged in its natural setting, but they learn to recognize reliable markings typical of the species. Experimental studies show that European kestrels locate voles by indirect signs. Voles are diurnal rodents that litter their runways with urine and fecal scent marks, and evidently kestrels can see the scent marks' ultraviolet (UV) wavelength.[5]

Experiments with striped skunks show how mammals may form search images based on scent. When they encountered a new kind of food several times, they got better at finding it and detected the scent up to ten times farther away than when they first encountered it. They formed olfactory search images even faster when the food reward was large. Strangely, however, the skunks lost their search image for a food when it was hidden in a different habitat: their search images seemed to be habitat-specific.[6] Learning search images undoubtedly gives many species an advantage in finding food more quickly.

AVERSIVE CONDITIONING: LESSONS WITH UNPLEASANT CONSEQUENCES

Animals learn to avoid stimuli or situations that have unpleasant consequences. The dog trainer's shock collar and choke chain are aversive conditioning tools, and dogs soon learn to heed commands rather than get shocked or jerked by the chain. The skunk's anal gland is an antipredator adaptation that operates on the principle of aversive conditioning (Figure 13.4). Naïve predators such as young bobcats and coyotes that tackle a skunk get a good skunking that they don't forget. The rapid acquisition of such awareness is called **one-trial learning**, and the skunk's unique coloration and odor are indelibly etched in the victim's memory.

But the lesson sometimes fails to stick. Even after a thorough skunking, some dogs apparently don't perceive skunks as a general class of threat. In the terms of learning theory, they fail to generalize the distinctive coloration and smell of their particular attacker to skunks in general. These dogs need multiple experiences to get the message that all animals that look, smell, and behave like a skunk are a dangerous menace. Some dogs never figure it out, and in these luckless victims the urge to attack seems to defeat the benefits gained from associative learning.

Wildlife biologists in Canada's Banff National Park used the principle of aversive conditioning to correct the behavior of tame elk that strolled through town like sacred cows, browsed in gardens, and caused traffic jams.[7] Hazing by dogs and people made them more wary of people and increased their flight distance, and the herd moved away from town, where they had become a nuisance.

Stimulus generalization may explain why coyotes recognize human scent as dangerous in different contexts. Wildlife biologists survey carnivores at scent stations, but one researcher found that previously trapped coyotes visited the stations less often than expected. It is likely that they associated human scent at the scent station with the negative experience of being trapped and immobilized.[8] They probably generalized the smell of people in the aversive experience to other situations that smelled of people.

Conditioned food (or taste) aversion is a special case of aversive learning that can be summed up simply: if it made you sick, don't eat it again (see "Lessons Learned by the Loner Wolves"). This kind of aversion is hard to document in nature, but the unpleasant manifestations of food poisoning in people and research animals tell us it resulted from natural selection. The eaters and the eaten, animals and plants, have been in an evolutionary arms race for a very long time, and most ecologists agree that plants and animals produce defensive chemicals to deter feeding by other animals.[9] Mammals and birds cope with toxic foods by taste-testing and selective feeding. When they eat toxic foods, they may dilute the poisonous effects by eating only small quantities. They may detoxify them metabolically in the gut. Or they may vomit.

LESSONS LEARNED BY THE LONER WOLVES

Aversive conditioning is an important tool in managing predators, but in the early 1900s it was an unintended consequence of the US Biological Survey's program to exterminate wolves from western rangelands. The government's rustic predator control agents and contract hunters trapped and shot wolves and exterminated them in their dens, but this was labor-intensive. A more expedient solution was to lace meat or livestock carcasses with strychnine (Figure 13.5). The program eventually succeeded, but only when the last cagey loners, christened by the agents with colorful monikers like frontier outlaws, were finally killed.

It is clear that those hounded fugitives learned to avoid their pursuers. They had survived sub-lethal strychnine poisoning and agonizing loss of toes in steel traps. Single-trial learning was probably one adaptive response to traumatic or near-death experiences, and conditioned food aversion is a plausible explanation for their sustained avoidance of traps and poisoned bait. The elusiveness of the loners was likely due to stimulus generalization, that is, shunning all odors and signs of human activity associated with previous painful misadventures.

Figure 13.5. A strychnine-laced horse carcass deployed by federal predator control agents in the early 1900s. All kinds of dead livestock were used. Wolves that survived poisoning evidently learned to avoid such carcasses. *Courtesy of the Smithsonian Institution Archives*

The flaw in the published accounts is the averred longevity of the loners, up to eighteen years. This raised a red flag for contemporary wildlife biologists, who examined the loner wolf skulls preserved at the National Museum of Natural History and found that modern aging techniques don't support the alleged longevity. It is likely that each putative loner was embodied in a succession of one or more doppelgängers.

Nonetheless, those hard-to-get wolves remain viable anecdotal examples of several forms of mammalian learning. Hunger probably trumped the loners' initial neophobia to human scent and steel traps, but the bite of the trap and the vile illness of poisoning taught them to avoid all traps and baits. Close calls with death and aversive conditioning didn't always give them stays of execution, but it certainly made them harder to catch.

Wildlife managers have used conditioned food aversion to train predators to discriminate between livestock or endangered species from other forms of prey. It is a difficult and time-consuming process because species with discriminating tastes, like pigs, can detect and reject the sickening emetic, or they may limit intake and avoid the effects of a full dose. Carnivores that "wolf" their food, like coyotes and wolves, are more vulnerable; but predatory birds, like ravens, crows, and jays, are less likely to reject treated baits because they have limited powers of taste.

Researchers tested captive Steller's jays to see if conditioned food aversion could lessen their predation on marbled murrelet eggs.[10] Jays hamper recovery of this endangered species that nests almost exclusively in old-growth conifers in the Pacific Northwest. Imitation murrelet eggs dosed with an emetic caused the captive jays short-term illness. The aversive effect lasted two months, the length of the experiment. Then they tested the effects on wild jays using treated murrelet-mimic eggs and untreated nonmimic eggs as controls. The jays preyed 37 percent less on the mimics than on control eggs.

Now let's examine the role of learning phenomena in the real-life melodrama of people and bears in the out-of-doors.

OF PEOPLE AND BEARS: LEARNING IN NATIONAL PARKS

National parks have taught us much about learning in wildlife and in people. A telling chapter in the ongoing saga concerns black bears, tourists, and the park managers of Yosemite.[11] Before the 1920s most bears in Yosemite kept their distance from people, and park staff and concessionaires tried to keep it that way. They hazed and chased bold and curious bears with dogs, and they shot the incorrigible. With no knowledge of learning theory, they succeeded in negatively conditioning bears to people.

This changed when park visits grew in the 1920s, and harmless encounters between bears and tourists led to mutual habituation. Park visitors then discovered the joys of feeding bears, which also rewarded the creatures for boldness and curiosity (Figure 13.6). At the time, increasing trash spawned the first garbage dumps, and the park's "feeding pits" became both theaters for bear watchers and moneymakers for the park concessionaire. Park naturalists cautioned that feeding bears corrupted their behavior and put bears and people in harm's way, but park managers and the concessionaire ignored the advice and defended the new order as good for business.

By 1930 Yosemite's tame bears had learned that where people gather, there was food. They foraged in campgrounds, picnic areas, and wherever their noses led them to human smells. They muscled their way into cars and food containers and bit and clawed the growing number of tourists who fed them. Irritated and maimed visitors were the

Figure 13.6. National Park visitors of the past fed bears and unwittingly reinforced bold and dangerous behavior.
Courtesy of the University of Wyoming, American Heritage Center, Laramie, Wyoming

unintended consequences of the liberal feeding policy, and bad publicity forced administrators to change their practices.

Administrators blamed bears for the problem and settled on a solution of physical removal and culling. Park staff captured problem bears and released them far from their temptations, but soon they learned that bears are wilderness navigators capable of homing back to familiar haunts. Repeat offenders were killed over the next fifty years. Removing curious bears amounted to an uninformed program of artificial selection, but the sad story of Yosemite's bears is punctuated with comic relief. Park managers and visitors had created a fertile landscape for the bears to explore and exercise their powers of problem-solving. The bears had uncanny abilities to circumvent obstacles and deterrents to food. With sheer strength and often the patience of Job, they opened food chests, garbage cans, and locked automobiles.

Among the ursine cast of characters were also a few risk-takers. Some bears were quite willing to take risks known only to stunt performers when park authorities advised hanging food bags on tree limbs as a deterrent to food theft. If the dangling prize couldn't be reached safely, some bears simply launched themselves at it like Flying Wallendas.

It also became clear that bears differed in inventiveness. Some became charming panhandlers that stood on their hind legs, flattered some park visitors into playing the animal trainer, and thus reaped more food than their less talented peers (Figure 13.7). Yosemite legend has it that a commanding matron gathered a crowd by hand-feeding a particularly cooperative bear. Approving onlookers were admiring her mettle when the bear's claws accidentally snagged her bodice. I'll leave the details to your imagination, but the lady's embarrassment was so great that she demanded the bear's permanent removal from the park. By this time in Yosemite's history, the authorities knew that the culprit was not the bear.

The craving for food also stirred the bears' cognitive powers and revealed the plasticity of their behavior. Many bears abandoned seasonal movements and settled in the valley. Others became camp followers and tracked backpackers into the high country beyond their previous ranges. Normally diurnal, they also shifted their activity to night, when human interference was less likely.

In the mid-1970s investigative reporters discovered that annual records of bear-killing in the park changed with the prevailing views of Yosemite's successive incumbent administrations.

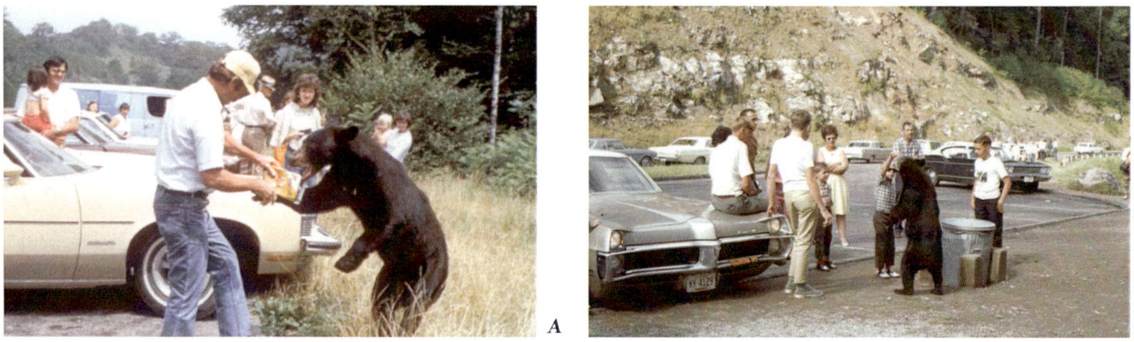

A B

Figure 13.7A–*B*. Some food-conditioned bears became charming panhandlers. (Great Smoky Mountains National Park, Tennessee)
A: *Courtesy of Gordon Burghardt;* ***B:*** *Credit: M. W. Meyers, National Park Service*

Some administrations were heavy-handed; others were not. The resulting exposé of managerial whim in handling problem bears caused a public outcry that forced management to embrace more biologically acceptable methods. Yosemite's administrators finally conceded that research might help solve the bear problem, and scientific studies of bear biology commenced.

Figure 13.8. Bears are entertaining until they show an interest in your car. (Great Smoky Mountains National Park, Tennessee)
Credit: NPS SVC #3337

One subsequent discovery was that the park's bears had discriminating interests in cars (Figure 13.8). Between 2001 and 2007, a research team tackled the problem of car break-ins and discovered that bears didn't break into just any vehicle they found. They chose minivans out of proportion to their availability and bothered sedans and sport utility vehicles (SUVs) less often than expected. In other words, they discriminated between makes and models of cars. The researchers offered several possible explanations.

One is that minivans are popular family vehicles, where parents distract irritable children with snacks. Spilled snacks become bear bait under and between car seats. Another possibility is that minivan owners are more likely to store food in their vans than in bear-proof containers supplied at campsites. Then again, a few clever bears may have learned that minivans were easier to break into than other cars. Though many bears lived in the study area, only a few bears became specialists in car break-ins. Evidently, these bears made the association between food and minivans and learned to discriminate them from other cars.[12] The shenanigans of black bears and tourists are an amusing lesson in comparative learning that inspired a ranger to remark: "There is considerable overlap between the intelligence of the smartest bears and the dumbest tourists."[13]

Yosemite's current management of food-conditioned bears is an improvement over past practice. The national park invested in visitor education, installed bear-proof lockers for food, and

A B

Figure 13.9. ***A***, Bear-proof food lockers are one way to prevent bears from becoming food-conditioned. ***B***, Live trapping and deportation to the backcountry and euthanasia have proven largely ineffective as bear management tools. Trapping, sedation, and euthanasia are undesirable but are often necessary solutions to problem bears. (Yosemite National Park) deported nuisance bears to the backcountry (Figure 13.9). However, aggressive bears, repeat offenders, and many deported bears that returned to their old haunts were euthanized. Bears paid with their lives for the carelessness of arrogant visitors, and managers decided to try aversive conditioning on animals that food-grubbed in campgrounds.

Nowadays, bears that venture into campgrounds find themselves confronted by a team of professional bear hazers who harass them with noisemakers and pelt them with rubber bullets and beanbags. Bears soon associate campgrounds with bad experiences and stay away. Avoidance behavior is the upshot of such aversive conditioning. Visitors who don't cache food in a bear-proof locker get a rude awakening when a stern ranger rouses them from sleep and advises them to clean up their campsite.[14]

Aversive conditioning is not a magic bullet. Biologists studied it in Sequoia National Park, where they classified bears as food-conditioned or wild. The difference is that food-conditioned bears were habituated to people and had an acquired taste for human food, while wild bears lacked those traits. Aversive treatments included chasing, pelting with rocks, spraying with bear repellent, or shooting with rubber bullets. They learned that bears with human-food dependencies were not easily conditioned, and yearling bears were more resistant to conditioning than adults or small cubs. Aversive conditioning worked best on bears that hadn't yet become hooked on human food.[15] As a management tool, aversive conditioning can reduce bear problems, but it can't eliminate them.

The problem is bears *and* people, because no amount of education or law enforcement is enough to correct human behavior that creates nuisance bears. Managing both will always be necessary in national parks and everywhere else they find one another.

CONCLUSIONS

Mammals and birds have amazing perceptual abilities, and their relationship to the environment is shaped by a variety of learning mechanisms. As camera trappers, we know that our subjects are aware of our visits and our equipment, but we don't dwell on how they think until they surprise us and do something unexpected. Usually, this happens when they tamper with our equipment, but when they go off-script in other ways the effect on our mindset can be spooky. Disbelief may turn into a surreal realization: that an anonymous subject is smarter than we thought and may even be keenly attuned to our activities. We can only conclude that they are more than just charming and beautiful subjects. They are watchful and intelligent in their own ways.

CHAPTER 14

Managing Your Photos and Data

Camera trapping generates a great deal of useful information and countless media files, which users manage in various ways to suit their needs. Essential needs are editing, storing, and retrieving media, backing up files, and protecting images on the internet from piracy. Annotating media files on spreadsheets is an optional activity that organizes and enhances the scientific value of your findings. Other useful databases include spreadsheets or map-based inventories of camera sets and tabulations of species and behavior recorded at each camera set. The internet offers several ways that our findings can be shared and used to benefit wildlife conservation.

Data management is an inescapable reality—a gratifying mental exercise for some, and drudgery for others. At minimum, camera trappers need a reliable way of storing images for quick retrieval, especially for large collections of photos and video clips. We may also want a handy record of all camera trap locations with related details. Perhaps we want to see our sets displayed on a topographic map or satellite image. Tools to manage camera trap data are readily available. Spreadsheets offer an easy way to tabulate data, and map-based programs like Google Earth give a bird's-eye view of camera sets. Trail camera manufacturers also provide programs and apps as aids to hunters. One, for example, helps understand deer movements in relation to weather conditions. Advances in image-recognition technology, artificial intelligence and machine learning have already been adopted to identify species and animals individually.[1] Web searches will certainly reveal a growing variety of data management programs and apps, along with reviews of their performance.

This chapter describes some basic forms of data management, from keeping track of cameras and imagery to summarizing biological observations in spreadsheets and summaries. The goal is to show what kinds of information can be used, with examples of how I have interpreted some findings.

MAINTAINING A TRAIL CAMERA INVENTORY

What do you do when your camera is stolen or stops working? Those who lead well-ordered lives know exactly where to find records needed for a police report, insurance claim, or service request. Those who file records like squirrels bury nuts—all over the place—rummage for the serial number, date of purchase, and vendor's name.

Making a spreadsheet is simple, and it means you can quickly find vital details about your cameras, whether you have few or many (Figure 14.1). You can also include certain operational features if you want to evaluate the performance of different makes and models at a glance. A column to note camera quirks and repairs is also helpful. If you own a fleet of trail cameras, use numbered codenames to tell them apart.

CODE NAME	MODEL #	SERIAL #	PURCHASE DATE	VENDOR	MODEL NAME
BR12	BTC-7FHD-PX	46100742041707PXZ	6/20/18	Minnovations	RECON FORCE FHD EXTREME
RESOLUTION STILL, VIDEO	**TRIGGER DELAY**	**SD CARD CAPACITY**	**DETECTION RANGE**	**FLASH RANGE**	**FOV**
20, MP, 1920 x 1080	0.4s	512GB	80 ft.	120 ft.	55 degrees

Figure 14.1. Sample line from a trail camera inventory spreadsheet presented with the columns stacked to save space and missing the final data field for Notes.

AN INVENTORY OF CAMERA SETS

Where did you take that picture? The answer is easy if you trap a small area, but you may draw a blank if you set cameras far and wide. A camera-set database is an index of your activities. If you want to revisit a particular site after a long absence, it tells you exactly where to go and what you did. It is easy to find other details if you keep cross-referenced folders of your sets and data tabulations. Tracking with this level of detail may sound off-putting, but the options described below make it easier to remember what you have done and find imagery.

GEOREFERENCED CAMERA-SET DATABASE

GPS waypoint numbers are a logical way to name camera sets and related folders and images. If you set a camera at one location repeatedly, identify successive sets with decimals ("Set 699.01"). If you have more than one camera at a location, add letters to distinguish them ("Set 699.01B"). Set numbers can be used to identify multiple images of the same subject. For example, a downsized image with a file name of "Arctic fox-824.12-172.jpg" tells you where to look for the original image (in the "Set 824.12" folder). The final three digits identify the specific image ("DSC00172.jpg").

Another option is to label your sets with a brief letter code. Abbreviations like "Qhr" for Quail Hollow Ranch and "Sar" for Sierra Azul Ridge are better cues to a location than numbers, and they are easier to remember.

A camera-set database preserves essential details like GPS coordinates, a description of the location, and other facts of your choosing (Figure 14.2). Fill in the data fields every time you make a new set. It's hard to remember camera sets by numbers, but when you have logged several hundred sets, you can still locate forgotten ones by searching keywords in the data fields. A camera-set inventory preserves a complete history of your camera trapping activities.

CAMERA SET #	699.01A
LATITUDE	61.66489
LONGITUDE	-145.174365
ALTITUDE (FT)	Unknown
COUNTY	Valdez-Cordova
STATE	Alaska
HABITAT	Aspen riparian/pond
LOCATION DESCRIPTION	Heavily used trail near bank, 75' from lodge
CAMERA ID	SLC-17
DATE SET	8/19/14
DATE CLOSED (OR RE-SET)	8/20/14
NUMBER OF CAMERA DAYS	1
NUMBER OF PHOTOS (TOTAL)	134
NUMBER OF ANIMAL PHOTOS	130
CAPTURE SUCCESS	97%
TIME FRAME	Night
MEDIUM	Stills
TYPE OF SET	beaver "skid trail"
CAMERA BEARING	WNW
DISTANCE TO SUBJECT	6'
ATTRACTANT	Castor
SPECIES	Beaver (130)

Figure 14.2. Possible data fields for a spreadsheet inventory of trail camera sets. The left column lists the data fields; the right column shows sample entries.

Mapping your camera sets is an alternative to a spreadsheet. Mapping software allows you to record camera locations visually in relation to geography and habitat, and you can include notes and photos. There are numerous programs for use on laptops, desktops, and even smartphones that display your data in various types of maps or overlays (see "Some GPS Mapping Software").

Google Earth is user-friendly, and online tutorials help to surmount most obstacles. Data entry is simple. Designate sets with the waypoint number, and then enter the longitude and latitude. This can be done manually for individual entries, or you can upload a series from your GPS unit[2] or a data table.[3] Google's Edit Placemark window can accommodate a large amount of text for your description of the set. You can also link your placemarks or waypoints to photos by clicking a waypoint number on the map to see photos of the set, the camera's FOV, or your favorite photo-capture. I mark camera sets with Google's camera icon. You can mark additional points of interest with customized icons to distinguish them from camera sets. Google Earth will probably satisfy your needs to map waypoints and tracks and help you plan excursions and explore landscapes. However, an internet connection is needed to access the global map.

RECORDING CAMERA-SET DATA IN THE FIELD

HIGH-TECH METHODS

Recording data in the field with a computer notepad, laptop, or cell phone is better than trying to remember details when you get home. You can enter data directly into a spreadsheet such as Microsoft Excel, Claris FileMaker Pro, or Quickbase. Or you can use a Google Form linked to a spreadsheet. Google Forms is free and easy to use; data can be entered on any desktop or mobile device and downloaded as a spreadsheet. It's easy to make simple data forms using older versions of Excel, but making flashy, user-friendly forms requires more skill, patience, and help from online tutorials.

A photo of the camera set and its general FOV is a useful supplement that jogs the memory better than a verbal description. FileMaker Pro allows you to import images, but presently there is no satisfactory way to display images in Excel. The best solution is to list **JPEG** file names in a designated column. Database programs are pricey, but most of us have Excel on Windows and Apple computers. Google Sheets or Apache Software's OpenOffice (https://www.openoffice.org/) are free, **open-source** alternatives. If your computer of choice is a cell phone, online tutorials describe how to access data forms using different cell phone operating systems.[4]

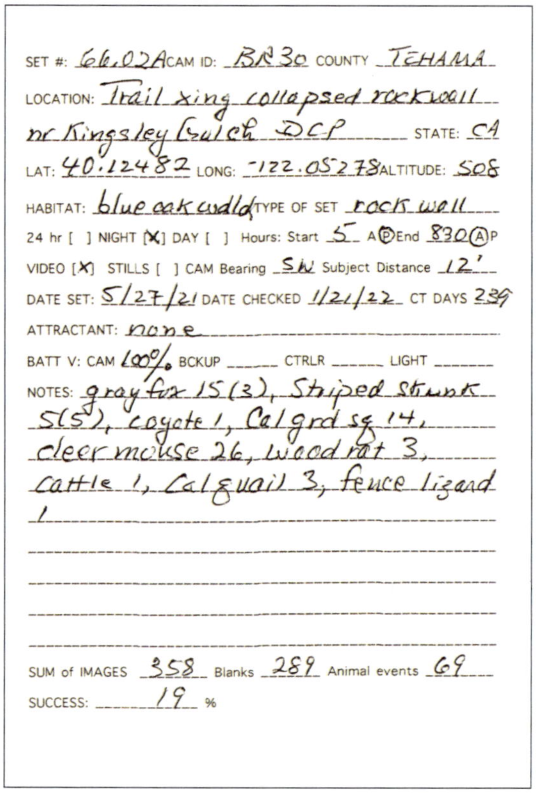
SET #: 66.02A CAM ID: BR30 COUNTY TEHAMA
LOCATION: Trail xing collapsed rockwall nr Kingsley Gulch DCP STATE: CA
LAT: 40.12482 LONG: -122.05278 ALTITUDE: 508
HABITAT: blue oak wdld TYPE OF SET rock wall
24 hr [ ] NIGHT [X] DAY [ ] Hours: Start 5 A Ⓟ End 830 Ⓐ P
VIDEO [X] STILLS [ ] CAM Bearing SW Subject Distance 12'
DATE SET: 5/27/21 DATE CHECKED 1/21/22 CT DAYS 239
ATTRACTANT: none
BATT V: CAM 100% BCKUP ______ CTRLR ______ LIGHT ______
NOTES: gray fox 15 (3), Striped skunk 5 (5), coyote 1, Cal grd sq 14, deer mouse 26, wood rat 3, cattle 1, Cal quail 3, fence lizard 1
SUM of IMAGES 358 Blanks 289 Animal events 69
SUCCESS: 19 %

Figure 14.3. Example of a completed camera trap logbook data form created with Microsoft Word.

SOME GPS MAPPING SOFTWARE

- **Google Earth** (https://earth.google.com/): worldwide satellite imagery with the interactive capacity to upload GPX files of GPS waypoints and tracks; see https://www.screencast.com/t/Q7PtVyPTfkfl for a tutorial about moving data from a GPS to Google Earth; see https://www.screencast.com/t/aUKyAeVC3cf for how to transfer waypoints and tracks from GPS receivers to your hard drive and then to Google Earth
- **Garmin mapping programs** (https://www.garmin.com/en-US/software/basecamp/): a complete two- to three-dimensional mapping facility for desktop and laptop computers (3D mapping available with a subscription)
- **iHike GPS** (https://www.ihikegps.com/): a map-based program for Apple mobile devices that allows import and export of GPX and KML waypoints and tracks to other devices and Google Earth
- **ArcGIS** (https://www.esri.com/en-us/arcgis/products/arcgis-online/overview): a powerful, professional, cloud-based mapping program that can be used online with a reasonable fee rate for individual users
- **QGIS** (https://www.qgis.org/en/site/about/index.html): a free, open-source GIS program usable on Windows and MacOSX machines for managing, editing, and analyzing mapped data
- **iNaturalist** (https://www.inaturalist.org/): a program for sharing geo-based, natural-history observations globally. Its crowd-sourcing function enlists specialists to help identify species; It's not a repository for all of your photos, but it's a good place to share special records and observations.

RETRO METHODS

Journaling with pen and notebook is an alternative to a spreadsheet. For consistent records, list your required data fields at the front of your notebook. Notebooks are simple and convenient. They also have space to sketch maps, record observations on animal sign and habitat, and note your expectations and experiences.

Printed data forms are another option (Figure 14.3). Draft your data form on computer, print multiple copies, and staple them together. You can carry them on a clipboard and file them in a folder for future reference. If you're handy, bind your own notebook. Format two data forms on a single sheet of printing paper and print them in landscape (horizontal) mode on both sides of the sheet. This yields four note pages per sheet. Fold them in half, and staple or stitch the folded pages together between soft cardboard covers. If you want a customized notebook without the work, take your layout to a printer and have them print a minimum order. And if you will be working in inclement weather, printing on Rite in the Rain paper is well worth it. Regular printer paper will disintegrate when wet. Rite in the Rain paper does not.

Notebooks may satisfy your needs as a primary source of records, but they require more work than a computerized spreadsheet. I used them for many years and updated each set with a list of photo-captured species and the number of photos of each. Thumbing through the pages, however, was not a satisfying way to search for records. It was easier to find particular sets after I filled the notebook and added an index page listing sets sequentially with corresponding pages.

HARVESTING AND PROCESSING THE CATCH

Many of us like to take a sneak peek of our images in the field, but it's on the home computer that we upload the photo-catch for careful viewing. If there's occasion for a happy dance, we hurry to share our pictures on social media. If you are inclined to view and study the catch, you might consider annotating the photo-captures for future reference.

TABULATING RESULTS FROM CAMERA SETS

Thoughtful review of the photo-catch is highly informative. Tabulating the results in a spreadsheet creates a permanent record of what transpired at the set, and it is a lot easier to review your results that way than poring through a folder of photos or video clips. Several programs are available for technical analysis of trail camera data,[5] but here I describe two options for general purposes.

OPTION 1: IMAGE ANALYSIS SOFTWARE

Timelapse (https://saul.cpsc.ucalgary.ca/timelapse/) is a free image-analysis program for field biologists who use camera traps for research. It reads and displays still and video media on Windows and Apple computers with windows emulators. It's quite a satisfying way to tabulate data. The tutorial is a quick and easy introduction to the program, and the user manual explains the many features. Timelapse's template is a way to customize data fields to your project-specific needs. It will then display the various fields for entering data. For example, you can easily create a drop-down list of species you are interested in.

Data are automatically stored in a CSV table, which can be viewed with the click of the mouse or imported for further analysis using Excel and similar spreadsheet programs. You can automatically scroll through images at various speeds, from two to ten frames per second, and stop when you see a pattern change. Or you can search for animal images by scanning a display of multiple thumbnails. A magnifier, zooming, and photo adjustments are handy aids to identify hard-to-see animals.

The manual describes Microsoft's process of classifying images as persons, vehicles, and animals. Timelapse can import the Microsoft recognition data and display it using color-coded bounding boxes. This is currently a free service and is an invaluable way of dealing with landslides of animal-empty images. The manual advises you to validate accuracy of the classification with the available tools. Timelapse includes many features for optimizing workflow, including various methods to rapidly enter data. You can also view and manage subsets of your images, zoom in on details, view and tag multiple images at the same time, and much more. Timelapse is a convenient and versatile alternative to annotating spreadsheets.

OPTION 2: JUST MAKE A SPREADSHEET

If you tabulate observations from camera sets directly into a spreadsheet, the result will be a sequential listing of photo-captures and any observations you wish to note (Figure 14.4).

SET 719.01, APLODONITIA TUNNEL UNDER ROOT OF FALLEN LOG, SF STATE'S SIERRA NEVADA FIELD CAMPUS, BASSETTS, SIERRA COUNTY, CA; "Sony-L"				
5/23/15 - 7/6/15 = 44 DAYS				
372 pics minus 201 blanks = 171 ANIMAL PICS; 46% SUCCESS				
Filename	**Date (YMD)**	**Time (HMS)**	**SPECIES**	**NOTES**
DSC00812.JPG	5/23/15	17:36:36	Shrew	climbing bank
DSC00820.JPG	5/26/15	00:07:51	Deer mouse	entering tunnel
DSC00987.JPG	6/20/15	10:16:33	Aplodon	in swarm of mossies
DSC00892.JPG	6/4/15	15:07:45	Chickaree	tail and rump only
DSC01015.JPG	6/20/15	15:16:36	Mosquitos	mossie squadron in focus
DSC01062.JPG	6/24/15	11:51:47	Aplodon	carrying corn lily leaf
			Mtn beaver	149 (19)
			Chickaree	11 (9)
			Unidd	2
			Deer mouse	6 (6)
			Shrew	3 (2)
			Animal Pics	171

Figure 14.4. Sample annotation of still photos recorded at a camera trap set. Normally, I do not delete blanks from the spreadsheet. The word *Aplodon* in the species column means mountain beaver. Photos per species and number of blanks were summed using Microsoft Excel's function keys (COUNTIF and COUNTBLANK). The values in parentheses are the number of visits (using a one-hour interval as the criterion).

Open a new spreadsheet, and then name the file with the set number. Reserve the top of the page for general information by merging several columns. Copy and paste the sequential list of photo files to the first column, and then download the corresponding date and time stamps in the second column. EXIF files containing date and time of each photo or clip can be downloaded in bulk from any folder containing photo files.[6] Many programs capture EXIF data for still photos, but you will need a program like GraphicConverter to batch-load EXIF data for both stills and videos (see "Correcting Incorrect Date and Time Stamps").[7] Use the remaining columns to record species identification, behavior, and anything else of interest, like molt, body condition, injuries, or identifying markings. You may want to highlight file names of exceptional photos and clips.

Review files to make corrections, refine comments, and mark unwanted photos or clips for deletion. I usually do this the next day, after I have had time to recover from the initial excitement or disappointment. The spreadsheet makes it easy to tally the number of species and the number of photos and visits of each. Calculate the photo-capture success of the set (number of animal photos divided by total number of photos taken) and add this to the summary at the top of the page. A spreadsheet summary is really just an index of camera-set results, so file it in the same folder with the media files. It's a time-saver when you are looking for specific details about a particular set.

PURGING UNWANTED IMAGERY

Tabulating data identifies media files you want and don't want, but whether to purge depends on the purpose of your photo collection. The conservative choice is to purge blanks but retain all animal images for possible future analyses. Otherwise, purge all blanks and image files of poor quality. You may also want to jettison the numerous clips of foraging gray squirrels and deer mice. Keep marginal photos of interesting subjects or intriguing behaviors. Even a poor photo of something unusual is worth saving. Once purged, your folder contains only the best of the catch.

CATALOGUING AND MANAGING IMAGERY

There are many ways to store media files. The preferred procedure is to upload files to a photo catalogue, image library, or cloud-based photo-sharing service like Google Photos. Photo catalogues or media organizers allow you to archive photos and videos and to search, sort, and filter the archive for media of interest. Finding desired files is simple when they are tagged with keywords. The tag of first order is almost always the name of the species. Additional tags enable refined searching. Behavioral tags, like foraging, drinking, bathing, feeding, playing, courtship, and grooming, help greatly when planning a video. Photo ratings save time in finding the best pictures and clips.

Let's consider three common methods for cataloguing and managing your image files.

METHOD 1: USE ADOBE LIGHTROOM

Adobe Lightroom includes photo editing and management modules. It has easy sharing ability, intelligent image search, and multiple fields for photo identification, including GPS waypoint and comments. It can also be installed on a cell phone. It is available by subscription, but if you cancel you lose the work you invested in the system.

METHOD 2: USE PROPRIETARY PHOTO-CATALOGUING SOFTWARE

Many photo-cataloguing programs, including the free software loaded on new cameras, let you tag photos with keywords, individually or in batches. Free media organizers sometimes accompany new cameras, but they usually don't recognize other makes of camera. They have limited usefulness if you use several kinds of homebrewed cameras.

METHOD 3: USE YOUR COMPUTER'S FINDER

File organizer apps on Microsoft and Apple computers allow searching for tagged files. Windows Explorer and Apple's Finder can search for keywords or tags in the file name, keyword, or tag field, and in the text of documents. You can tag a nondescript file like "DSC_0061.mp4" with a keyword, or you can simply rename the file as "Puma-886.01-61.mp4." Both systems will locate files when you search for "puma." They will display EXIF values and other metadata for all files containing the word *puma*, and they will list all files containing that word, including documents and PDFs. These apps don't have the versatility of a photographic catalogue, but they will serve you well if you faithfully tag your media files.

CORRECTING INCORRECT DATE AND TIME STAMPS

Finding that your date and time stamps are wrong could be a lament from the camera trapper blues. Anyone can inadvertently reset their camera to the default date and time when changing batteries. Computers may also change date stamps when you copy or move media files between devices. These are good reasons to check camera settings often and to preserve original files as though they were sacred reliquaries.

Photo-cataloguing programs, inexpensive utilities, and several applications let you bulk-correct misdated files. You can also correct data in Excel if you have the presence of mind to check the date and time settings while in the field. Trigger a final trail camera photo and write down the correct day and time. Then download the EXIF data and subtract the difference between the last incorrect date and the correct date: DATEDIF(D3,D4,"D"). The correct date would be the day you recorded the camera's date and time. Add that difference to the incorrect values in the other cells. Unfortunately, this method doesn't work if the date cell also contains time data.

EDITING STILLS AND VIDEOS

Free and inexpensive photo editing programs provide most of us with suitable editing tools. Camera trappers usually crop, downsize, color-correct, and sharpen images, and they often add copyright stamps before sharing still images on social media. Advanced programs like Adobe Photoshop and Adobe Lightroom offer additional tools, but mastering their full capabilities takes time. How-to manuals, YouTube videos, and online courses can ease the way.

Most programs forgive mistakes, and you can almost always undo edits, but some actions, like overwriting edits to original images, are irreversible. Save your photo edits with a new file name. Some of my favorite early photo-captures exist only as poorly edited and downsized copies because I saved my edits to the original file.

Prices of video-editing programs mirror the complexity of the program, but free video editors have many useful features. You can drag and drop selected clips to arrange sequences and trim, split, and merge clips. You can freeze frames, insert still images, and adjust speed for slow-motion and time-lapse effects. You can also tune color and sound and add voice-over, music, titles, and transitions. Video editors differ in ease of use, availability of technical support, and ability to convert multiple video formats. The latter is important if your trail cameras use different video formats or outdated ones, like AVI and AFS. Camera trappers are usually satisfied with free video editors, and when they encounter a limitation, buying the commercial version usually solves the problem.

Advanced video editors have powerful editing abilities and special effects. They process HD videos generated by DSLRs, action cameras, and smartphones, but they perform best on fast new computers that can render a finished video for Vimeo or YouTube without waiting for hours. They are superior to inexpensive editors for organizing large numbers of clips, and they offer new special effects that can greatly enhance the visual experience.

The multicam editing feature, for example, improves a scene's perspective by easily incorporating alternative angles from multiple cameras. The motion tracking tool is ideal to highlight the mechanics of high-speed running in mammals by highlighting the changing curvature of the spine or length of the stride. Special effects are no substitute for good footage of animals doing interesting things, but when camera trappers use the power of advanced video editors with excellent footage, their wildlife videos will look truly professional.

Research the options before selecting an editor. There is no shortage of reviews on the internet, and with a free trial you can test editors of particular interest. Printed manuals like the *For Dummies* series (Wiley Publishing) and YouTube videos can help greatly in learning to use video editors.

VIDEO FILES

SUMMARY MOVIES

It takes time to view video clips from highly productive sets, but you can view the action without the boring parts by making a summary movie. You can make separate summary movies for each species of interest or show the complete flow of events with one movie featuring the full cast of characters. Trim your best clips (remove blank segments) and merge them into a single file. The result will be an action-packed sequence showing patterns of animal movement and differences among species.

A summary makes it easier to select the best clips for a movie. It may also inform you of the need to change your camera position for different and better angles. Some summaries tell an interesting story by themselves. Sequences that span several seasons can be evocative, especially when peppered with transitional clips of seasonal change. Autumn winds, long shadows on short winter days, and drifting and melting snow add context and mood to the procession of wildlife.

CLIP CATALOGUE

You might want to catalogue copies of your best clips for reference and future moviemaking. After trimming and converting the clips to the desired format with your editor, label them by species and activity. "Bear rubbing tree-820.16-37.mov" identifies the species and activity, the set, the clip, and the video format. Sort the edited clips into species folders. As clips accumulate, make subfolders for different topics. If you assign keywords to every clip of value, you can always find the original files.

IMAGE PROTECTION

Millions of people post photos and videos on social media and download images and video for personal use. It is not illegal to share digital images on hosted sites, but it is illegal to copy and sell images from the internet without permission. There are ways to reduce the risk of piracy. If you copyright your photos, you have legal recourse against infringement.

The Creative Commons (https://creativecommons.org/), for example, offers bloggers and users of other internet platforms several options of control over their photographs and texts.[8] Users can choose a license that suits their interests; each type of license grants different rights. At one end of the spectrum is the public domain license that waives all copyright interests. At the other end is the familiar all rights reserved license. Between these extremes are licenses that permit sharing of content to different degrees with regard to attribution and commercial use.

There are other ways to protect photographic products. Labeling your photos with the copyright symbol, your name, and the year is a statement that the law protects your work. A large watermark label across the image is a stark declaration of ownership. You can also claim ownership in the IPTC metadata,[9] which is not visible on the photo itself. It allows the photographer to label the image with detailed information about content, copyright, and use. You can also specifically designate copyright status (public domain, unknown, copyrighted) and make a copyright notice. Here's an example:

> I, John Doe, am the sole owner of this photographic image, which is protected by United States and international copyright laws. Use of this image for commercial purposes is strictly prohibited unless I specifically grant permission in writing.

You can also title and caption the photo and identify yourself as the author. Filling in all the information can be time-consuming, but you can batch-label multiple photos using Photoshop, Lightroom, and other programs.

Images and clips that lack metadata are essentially anonymous because their owner cannot be located.[10] You can register your photos for a fee with the US Copyright Office online or by mail, which will protect your photos for seventy years after your death. If you publish videos on YouTube or Vimeo, they are automatically copyrighted to your name. People can still erase or change the data, but if you have to go to court, you can show that embedding copyright data is your established practice. If you find an unauthorized copy of your photo on the internet, you can send a *takedown letter* to the webhosting company.[11] It is easier to discourage misuse by posting downsized files. Pirates can't satisfactorily restore resolution of files with a resolution of 700 KB or less.

BACKING UP FILES

You are living dangerously if you don't back up your programs and files. If your computer crashes, you may be able to recover some or all of your images for a fee. Avert disaster by backing up files to a couple of external hard drives. All computers can be set to do this automatically. Since hard drives can also fail, computer specialists advise using two or even three backup drives and storing the third drive at the house of a friend or family member. Or store your photos in the cloud via DropBox, Google Drive, or OneDrive. If your hard drive fails, your data are still safe in the cloud drive.

PRESERVING THE PAST FOR THE FUTURE

Wildlife photographs can tell us about the past. Some camera trappers want to preserve their photo archives for the historical record. Why not? George Shiras's precious photographs are still accessible at the National Geographic Society, and his book *Hunting Wildlife with Camera and Flashlight* remains an enduring window of North America in wilder times.

Victims of hard-drive crashes may not agree that digitized media will outlast ephemeral media like books, photographic negatives, and prints, but digitization is here to stay. The Project Gutenberg online library was inspired by the premise that "anything that can be entered into a computer can be reproduced indefinitely." Now thousands of digitized out-of-print and copyright-free books are accessible globally.[12]

Our trail cameras record wildlife on a planet subject to climate change, development, and invasive species, and sometimes our video clips record rarely seen or unknown behavior. Donating documented media files to archives, posting and publishing video findings, and rendering trail camera services to scientific projects can help conserve nature.

iNaturalist.org, or iNat, is a repository of wildlife observations, documented with digital images, videos, and sound recordings that helps scientists understand species distribution, seasonal movements, and how populations change over time.[13] It is a reference library of data and expertise for curious naturalists to tap into when they can't identify plants or animals or just wish to share observations. Designed for naturalists of all stripes, it is also a social network for sharing discoveries and learning about life forms encountered in nature. Registered contributors submit photos and sound recordings of plants and animals with location and time of observation, and when specialists validate the identification, the record is added to a global biodiversity database. Camera trappers can archive noteworthy photographic records of species occurrence in iNat for posterity. The iNat site does not currently support uploading video, but you can link a screenshot in your iNat observation so your video can be seen at its hosting site, be it Flickr, YouTube, Vimeo, or Google Photos.

Contributors can easily access a list of their contributions, and when they have amassed fifty verified records, they can draw up online projects with specific goals. In short, iNat is an ever-growing database and powerful learning tool, but it is not an archiving solution for the rare folks who have meticulously documented collections of wildlife photographs and videos for decades.

Wildlifeinsights.org, like iNat, is a cloud-based platform and global database for biologists, universities, agencies, and NGOs that use camera traps to document and monitor mammal and bird populations. Its conservation purpose—to share, manage, and analyze data sets—thus offers recreational camera trappers a way to contribute their data to a greater cause. Service fees apply to organizations but can be nominal or waived for other users. Approved data contributors can upload images for species identification and analysis. An AI model has been entrained to identify nearly three thousand species of birds and mammals, and a future goal is to identify species from sounds and video, and perhaps individual animals. Both iNat and Wildlife Insights seek to conserve biodiversity by increasing awareness, but they differ in taxonomic scope, methodology, and contributors.

Camera trappers may also inquire of museums and professional biological societies for interest and advice in archiving long-term data. The bioacoustic or natural sound archives of Cornell University and Ohio State University are examples of how audio recordings are accepted on evidence of provenance and species identification. Hopefully, professional biological societies will consider their example in planning trail camera archives and developing vetting criteria. The American Society of Mammalogists created a photographic library for educational purposes decades ago and seeks photos of species not in the library, as well as photos of morphological traits and special behavior.[14] Other institutions, like university-based museums, might be interested in archiving documented research-quality collections of photos of scientific value.

Social media platforms don't guarantee perpetual access to their contributors' works, but they are a popular way to post and disseminate short wildlife movies and video clips. Posting on YouTube and Vimeo is a form of publication without editorial review, but contributors of interesting wildlife movies often disclose little or nothing about location and circumstances. Camera trappers give their work a measure of scientific validity when they provide those critical details.

We can also contribute to science by volunteering with federal and state agencies and with land trusts like the Nature Conservancy. The National Wildlife Refuge System and the National Park Service engage citizen scientists to meet their research goals, especially regarding the consequences of climate change on habitats and wildlife.[15] Land managers need camera trap data on a variety of conservation issues and benefit from volunteers helping on team initiatives.

Hard-drive crashes will remain an ever-present risk for camera trappers who don't back up photo files on multiple hard drives or store their photos on cloud-based services like Google Photos. Personal hard drives are apparently built for a finite lifetime. Cloud-based storage is safer because providers maintain multiple copies, but it doesn't make that nagging question disappear: Who will be the custodian of your archives when you are gone?

For most camera trappers, the old adage rings true: "It's not the destination. It's the journey." Few of us deny the loss of nature that we are witnessing. Sharing enthusiasm for the journey engenders concern for the natural world. The additional steps described in this chapter can also help; they are just different ruts in the same trail.

CHAPTER 15

Learning from Your Data

> Learning from experience is inevitable, but learning from your own data can be a gratifying reward of camera trapping. Photo records, when analyzed, can reveal the wealth of information buried in your photo files. This chapter discusses differences between observational and experimental data and how to convert raw data to camera trap days and animal detections. I also share some of my trials while seeking answers to questions about wildlife. The examples show that sorting and graphing observational data helps us visualize patterns of animal behavior, raises new questions, and sometimes surprises us with new insights.

Shy Boy was an amusing outlier of bear behavior. I discovered him on a video clip as he encountered my trail camera. He froze in his tracks, rolled his eyes suspiciously, and bolted. Had I camera-trapped this endearing bruin in the past? I scoured my photo folders for other clips,

Figure 15.1A., The "full monty." When back-rubbing on tree trunks, bears reveal their sex and often their unique breast blazes. (Butte County, California; video screenshots, Browning Recon Force Extreme)

Figure 15.1*B*. The Gallery of Rogues. My trail cameras also revealed occasional blindness and distinctive head scars in many older bears, especially males. (Butte County, California; video screenshots, Browning Recon Force Extreme)

and soon I was happily sorting mug shots and body portraits of bears. Those old photo archives were a treasure trove of forgotten information.

I compiled a spreadsheet inventory of bears and noted coat color, distinctive markings, body condition, and, where possible, sex, age class, and whether sows had young. Some of them had white blazes on their breasts, but the marks were hard to see. That prompted me to set cameras at their rubbing trees, where they bare their blazes and other personal details to the camera (Figure 15.1*A*).

The cast of characters included a gallery of rogues (Figure 15.1*B*). Scars of misadventure marked the gnarly seniors. Two had broken noses, and a few veterans had tooth-pocked scalps and scarred faces. Three old bruins hobbled on broken legs, and three younger animals were blind in one eye. As the database grew larger, it was hard to remember who was who. I added a column in my spreadsheet for names. I did find a few more clips of Shy Boy, but he disappeared forever that fall.

Pouring through the photo files was a satisfying digression. I improved at sexing and aging bears. I learned that coat color is a fickle trait that changes with molt and lighting conditions. I learned that many bears just pass through my camera trapping area and that residents usually disappear after three or four years at most.

DATA PROVED IT WAS A LEMON

I needed numbers to convince a trail camera company that a new camera was a lemon. I told the service representative that I had a dozen cameras of this model, but this one was setting new records for blank photos. He responded that blank photos are a fact of life, but I pressed my case and he agreed to take a look. The camera was back in my hands within a month with a note: "The camera was tested in all modes with no issues." The burden of proof was on me, but I needed data. So, I tested my lemon next to two normal cameras of the same model. It consistently recorded about five times more false triggers than the normal cameras. I wrote another letter and included the data and camera, which evidently proved my case. This time the company sent me a new camera.

BECOMING A DATA FREAK

Camera trapping is observing nature and accumulating photographic records. Inevitably, we form impressions about animal behavior and ecology. If we take the trouble to convert our images into data, we can test hypotheses and gain further insights. Biologists are doubting Thomases by training. They like to challenge opinions: "Do you have data to back that up?" and "Show me the graph!" Analysis feeds the healthy skepticism that comes with scientific thinking (see "Data Proved It was a Lemon").

You don't need a college degree to analyze camera trap data. Grade-schoolers can master analytical concepts and skills, and anyone can learn to graph data to visualize patterns. With tutoring you can even perform simple statistical analyses. If you want to explore the many ways that trail cameras have been used to test hypotheses, Kucera and Barrett's paper "A History of Camera Trapping" gives a wide range of examples.[1]

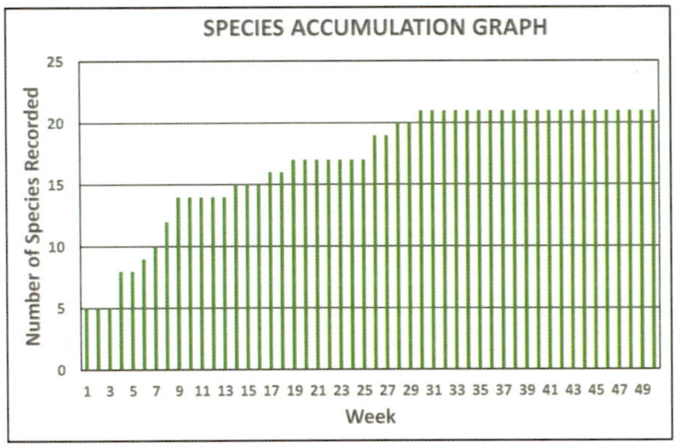

Figure 15.2. Graph detailing cumulative numbers of bird, mammal, and reptile species detected over four years at Audubon Canyon Ranch, Marin County, California. Many species were not detected, and intensive sampling of microhabitats would be necessary for a more thorough survey.

This chapter demonstrates how manipulating and graphing data helps to visualize patterns, answer some questions, and raise others. Observational data are a useful by-product of camera trapping, but they are often messy. It's preferable to design experiments to test specific hypotheses, but I suspect that recreational camera trappers are not interested in experimentation. Nonetheless, observations of wildlife often raise questions, and my goal is to introduce some ways of answering them. The first step is to prepare data for analysis. Readers who have little familiarity with statistics may find the following material daunting or uninteresting. Please steel yourselves, and do not feel bad if it is a struggle to follow.

TWO IMPORTANT MEASURES: SAMPLING EFFORT AND DETECTION EVENTS

CAMERA EFFORT

Trail cameras are sampling devices. Time-stamped images tell you when photos were triggered. They also record camera trapping or sampling effort. The basic unit of sampling effort is the twenty-four-hour camera trap day, or camera day. For example, three cameras set for three months (3 cameras × 90 days) photo-capture wildlife for 270 camera days. It takes one camera nine months to achieve the same sampling effort. If you plan to analyze data, set all cameras for the same sampling mode.

Calculating sampling effort helped me and my colleagues find an elusive species. Spotted skunks failed to show up in our camera trap survey of the Chimineas Ranch, but we knew they occurred in the county from museum specimens listed in the online Arctos database. Puzzled by our lack of success, we searched the literature and learned that spotted skunks prefer closed-canopy habitats. When we calculated camera trap days in each of the six habitats on the ranch, we learned that we had under-sampled the closed-canopy habitat known as chamise scrubland. There we deployed additional cameras, and there we recorded spotted skunks.

Camera trapping effort (more sites and more time) correlates positively with the number of species recorded, proving that the more you try, the more species you record. A practical example is the species accumulation graph. You can track your progress in surveying species by plotting the cumulative number of species against time (in weeks of effort) or in camera days (Figure 15.2). Accumulation curves are usually steep at first and level off when you have photographed most of the detectable species possible for an area. In theory, you could photo-capture every species of mammal known to occur in the area, but in practice this doesn't happen. Even when surveying a subset of mammals, like carnivores, certain species defy easy detection because of secretive habits or specific habitat preferences (see "Recognizing Habitat" in chapter 6).

DETECTION EVENTS

Biologists often use an interval of time, such as one hour, as a criterion to define independent visits or **detection events** of their photo-captured subjects.[2] This helps avoid double counting one animal that walks back and forth in front of the camera multiple times in a short period. Species with distinctive markings can be identified as individuals, but most mammals cannot be. Detection criteria make it possible to analyze data on populations of anonymous individuals. Multiple photos of a species are considered a single detection event when the photos are separated by intervals of less than an hour. This convention reduces high counts of animals that linger in the area. By calculating detection events, we can examine patterns of visitation over twenty-four hours and in different seasons and habitats.

Consider a camera set that recorded twenty-five photos of deer and twenty-five photos of striped skunks. The deer were photographed over twenty minutes; the skunk pictures were spread out over a month. When we apply the one-hour criterion, we find that identical numbers do not mean the same thing. The deer photos were taken during one visit, or detection event, whereas eighteen of the skunk photos were single photos taken on separate nights. The other seven were three detection events of two, two, and three photos each. Thus, the twenty-five skunk photos represent twenty-one detection events, or visits.

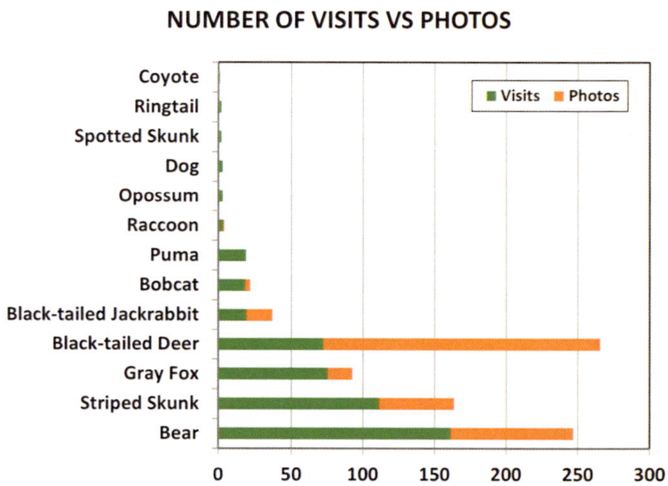

Figure 15.3. The number of photos may not equal the number of visits (or detection events). The green bars show the number of visits, while the orange bars show the number of photos that exceeded visits (redundant photos). Where orange bars are missing, the number of visits and the number of photos are the same.

Figure 15.3 compares visitation of thirteen species on a trail near my home over a continuous period of 781 camera trap days. Green bars show the number of visits based on the one-hour criterion, and orange bars show the number of photos that exceeded the number of visits. Notice that seven species lack orange bars, because the number of photos equaled the number of visits. It's tempting to conclude that bears are more common than gray foxes, but I can identify most of the local bears, and I know that a few animals accounted for most visits. The take-home message is that detection events, as defined, reflect animal traffic and not population size.

PHOTO-CAPTURE SUCCESS

Photo-capture success is another useful calculation, measured by dividing the number of animal photos by the total number of photos. The disappointing value of 33 percent tells you at a glance that only one in three photos are of animals, and that most of your photos were blanks. This measure lets you compare camera sets of different duration. If you calculate capture success by species, you can compare your success at detecting different species. The resulting visits, or detections per day, can also be compared between habitats or seasons. Species-detection success is usually less than one, because animals rarely visit a camera every day.

FINDING PATTERNS AND TESTING HYPOTHESES

Hypotheses are simply ways of asking questions. Biologists test hypotheses to see how animals respond to different conditions or experimental treatments. They state them in two ways—null hypothesis and alternative hypothesis. The null hypothesis is the prediction of "no difference." For example, we could say there's no difference in the number of times that bears come and go on trail X. We could also phrase it as bears travel in both directions equally on trail X. The alternative hypothesis predicts the outcome based on evidence or logic.

Let's look at some common questions and how examining the data gives insights or leads to answers.

Scientists design experimental studies in the laboratory, where they can control conditions and randomly select subjects for statistical analysis. Getting an adequate sample size is not a problem, and they can present the animals (or subjects) with different conditions, which are called *independent variables*. The subjects' responses are called *dependent variables* because they are expected to depend on the conditions.

Camera trap studies can be observational or experimental. In observational studies, researchers look for patterns and associations under natural conditions, but they have limited ability to control conditions (or **variables**) affecting the animals (or subjects). For example, we can use observational data to test the hypothesis that temperature is positively correlated with duration of baths in bears. The null hypothesis would predict no difference in bath length at different temperatures. The independent variable would be air temperature, and the dependent variable, measured from our video clips, would be duration of bath. If we camera-trap a small area, we will probably get repeated observations of a small number of animals. These are not independent observations; they are sometimes called *related samples*, and they are more prone to give biased measures. For more accurate statistical estimates, we should collect data from a large number of individually identifiable animals.

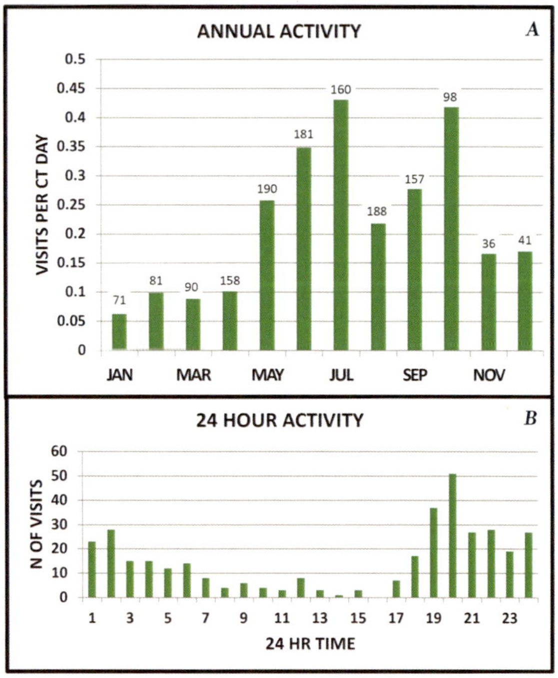

Figure 15.4. Annual and 24-hour activity of gray foxes based on eight years of data recorded at a creek bed in Butte County, California. ***A***, the bar height was calculated by dividing the number of visits by the number of camera trap days noted above each bar; ***B***, the graph plots the number of visits per hour ($N = 360$ visits). Data from different seasons were pooled.

If we wanted to test for age and sex differences in bath length, we would sort our data on bath length by sex into separate treatment groups. If we find no significant difference in the mean bath length between sexes using the **t-test**, we could pool the data for both sexes. If we plotted length of bath against air temperature, we would expect bath length to be positively correlated with air temperature. We might find that the slope of the correlation differs between sexes. Another approach would be to compare mean bath length at lower temperatures (say, 65–80 degrees) versus high temperatures (> 90 degrees).

Designing controlled experiments in the field is difficult because nature doesn't like to be controlled. Observational studies are an alternative way to study a species' occurrence or a subject's behavior under naturally occurring conditions. In the following paragraphs, I describe some examples of how I used camera trap data to answer a few questions.

WHEN IS MY FAVORITE SPECIES ACTIVE?

Graphing activity data helps to visualize twenty-four-hour and annual activity cycles. Figure 15.4*A* shows the monthly activity of gray foxes at a streambed used as a travel route. Foxes there are often photographed in spring and summer when they drink or forage for crane flies and moths, but they visit the site infrequently from November through April when the creek flows heavily. The low number of camera trap days tells you that I often pulled the camera to escape floods. The dip in visits in August and September and the second peak in October are hard to explain. Since this water hole does not go dry, the dip could result from a windfall of food in another area, like apples and manzanita berries. We could test that **hypothesis** by setting cameras in those areas.

The daily activity cycle in the graph in Figure 15.4*B* shows that gray foxes are active day and night, though 83 percent of visits happened between 6:00 a.m. and 6:00 p.m. My impression is that daytime activity is more likely in spring when pairs are rearing their pups. Peak activity was at 8:00 p.m. No activity was recorded at 4:00 p.m. Pooling data over multiple years can mask seasonal differences related to photoperiod (length of daylight) and availability of food and water. This is especially true when climate varies strongly from year to year.

DOES TRAIL TRAFFIC FLOW BOTH WAYS EQUALLY?

One of my favorite locations is a busy game trail where I once set a camera for head-on videos of bears and pumas. I expected them to move equally in both directions. After viewing several months of clips, I got the impression that more bears were going east than coming west. But when I tallied

Species	West	East	Total	One-Tailed Probability
Striped Skunk	27	59	86	<0.001
Gray Fox	13	47	60	<0.001
Black Bear	77	72	149	0.37
Bobcat	8	9	17	0.50
Puma	8	12	20	0.25

Figure 15.5. Direction of movement of five species on a game trail in Butte County, California.

the traffic data for the five common species, I got a surprise (Figure 15.5). My hunch was wrong for bears but correct for skunks and foxes. Bears, bobcats, and pumas traveled east and west more or less equally, but striped skunks and foxes traveled east 69 percent and 78 percent of the time, respectively.

You can't always trust your hunches. A colleague informed me that the one-way traffic habit in striped skunks was described many years ago, though it isn't that way on all trails.[3]

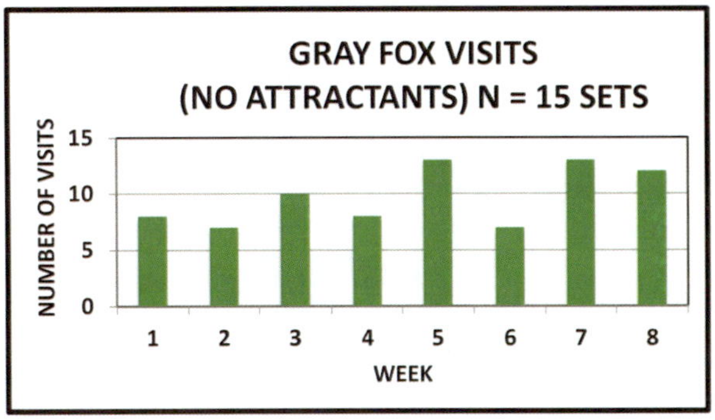

Figure 15.6. Gray fox visits to sets lacking attractants during an eight-week period. Data were pooled from fifteen sets in five northern California counties. Eighty-four camera trap days were recorded each week.

HOW DO ATTRACTANTS AFFECT VISITATION?

Cameras set at baits and scent lures attract more species and take more photos per detection than sets lacking attractants.[4] Figure 15.6 shows the steady visitation rate of gray foxes to camera sets on trails when no scent lures were used. Now look at the gray fox graph in Figure 15.7. The rate of visitation waned over time. These data were collected at the Chimineas Ranch, where we "doctored" our sets with scent lures and punctured cans of fish.[5] During the first four weeks, foxes visited scented sets more often than the unscented sets in northern California. Biologists would agree that these data are "suggestive," which is a polite way of

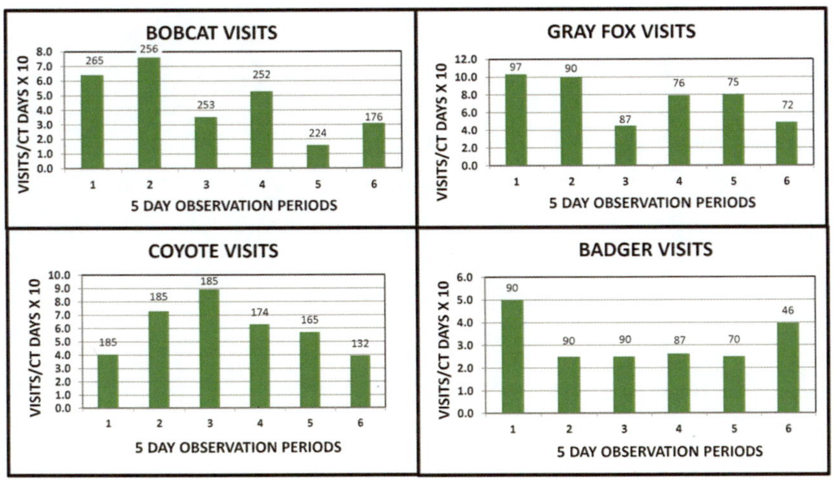

Figure 15.7. Visits of four carnivores to scented camera trap sets during six successive weeks at Chimineas Ranch, San Luis Obispo County, California. Species data were pooled from multiple sets: gray fox (22), badger (18), coyote (57), and bobcat (53).

Species	Number of Sets	Mean	Standard Deviation	Median	Range
Gray Fox	22	7.9	7.3	6.0	1–31
Bobcat	50	11.0	10.2	7.5	1–37
Coyote	63	10.9	149	0.37	1–39
Badger	18	18.3	17	0.50	1–39

Figure 15.8. Day of first visit to scented camera trap sets by four species of carnivores. (Chimineas Ranch, San Luis Obispo County, California)

saying you may be on to something, but you haven't proved it. Pooling observational data from different locations can cloud results, because a species' physiology and environmental conditions differ geographically and change seasonally.

Biologists design experiments to control this kind of variability. Had I planned this as an experiment, I would have set twenty pairs of cameras with and without only one kind of scent lure. Those lacking scent comprise the **control group**. I would have spaced cameras two miles apart to maximize the chance of recording different animals. I would have set all the cameras in one habitat, like chaparral, for the same amount of time, like eight weeks. I would have also made sure that my batteries and memory cards had the capacity to record lots of activity. When cameras fail in an experiment, the number of camera trap days differs and more calculations are in order.

The declining visitation pattern of the gray fox made sense. Attraction wanes as gamey fragrances fade with time. What about the other three carnivores at the ranch? Figure 15.7 shows that the bobcat and the coyote shared waning visitation profiles with the gray fox, but the badger's twin-peaks pattern is harder to explain. Maybe it reflects a low abundance of badgers and a reduced likelihood of encountering the camera traps.

Perhaps the coyote graph has you wondering: Is peak visitation in the second week a statistical artifact? Or did camera-shy coyotes delay their first visit despite the enticing scent? In the previous graphs, I plotted all visits of each species, which undoubtedly included visits of multiple animals and an unknown number of repeated visits.

Next, I wanted to tweak the data and look at only the number of days to the first or initial visits at each camera set. We can call this the **latency to first detection**. Some first visits took place from two to five times on the same day, and I suspected this was due to individual repeaters or family groups. So, I counted them as single visits to eliminate inflated values. Figure 15.8 summarizes the data set for each species.

All species made first visits as early as day 1 and as late as day 31 through 39 (weeks 5 and 6), as shown in the range column at the right. The mean day of first visits shows that the gray fox was the early bird, arriving on average 8 days into the set. The badger was last to arrive, with a mean date of 18.3 days. Bobcats and coyotes were intermediate, with an average first appearance on day 11.

The standard deviations were large, meaning first-visit days did not accumulate near the average value of a bell-shaped curve. The median statistic, which is the exact midpoint in the distribution of first-visit days, shows that bobcats were earlier (7.5 days) than coyotes (10 days). I presume that the coyotes' first visitation peak in the second week contributed to this. Different factors can influence this latency to first detection, such as species density in the area and behavior. Behavior can also change seasonally.

Species	CT Days	N of Detections	Detections per day	1-*p*	ln(1-*p*)	2.99573/ ln(1-*p*)
Gray Fox	739	84	0.11	0.89	-0.12	24.8
Bobcat	1792	163	0.09	0.91	-0.10	31.4
Coyote	1190	136	0.11	0.89	-0.12	24.7
Badger	547	33	0.06	0.94	-0.06	48.1

Figure 15.9. Example calculation of the number of camera trap days needed to photo-capture four species with 95 percent confidence using camera sets at Chimineas Ranch, San Luis Obispo County, California. (Based on the method used by Tobler et al., "Evaluation of Camera Traps," 11(3): 169–178.)

I also tabulated first dates in the twenty-seven sets that were visited by both bobcats and coyotes. Bobcats were first to arrive 59.3 percent of the time. The difference between the means was not significant: the two-tailed *P* value equaled 0.9691.

These **descriptive statistics** clarified my vague impressions. In all species some animals initially visited scented sets as early as day 1 or as late as several weeks after the set was made. Gray foxes were usually first to visit, and badgers were last. First visits of coyotes and bobcats were intermediate and similar.

Finally, let's compare capture probabilities of these four species. This statistic calculates the number (n) of days necessary to register a species with 95 percent confidence. In Figure 15.9, n is the total number of camera trap days, p is the average, or mean, number of photos per camera trap day, and 0.05 is the 5 percent probability of not capturing the species at all.[6]

The raw data are the camera trap days and the number of detections (or visits) for the four species (columns 2–4). We calculate detections per day (column 4) by dividing number of detections by camera trap days. To calculate 1-p, simply subtract detections per day from 1. To find the logarithm of that value, that is, ln(1-p), in column 6, go to an online logarithm calculator, select $\log^e$, and type in the 1-p value. To calculate your capture probability (column 7), divide ln(0.05), which equals 2.99573, by the ln(1-p) values in column 6. This yields the number of days expected to get the first detection within a 95 percent confidence limit.

Values of latency of detection are based on the day of the first visit. In all four species, they are greater than the values based on the frequency of all visits. The requirement for 95 percent confidence also contributes to the larger number of days. Notice differences between species. In this analysis, the probability of getting gray foxes and coyotes is the same, at 25 days or 3.5 weeks, but bobcats cannot be expected with 95 percent confidence sooner than a month. The probable time of first detecting badgers is 38 days, which is almost three times the mean delay to first visitation.

This was one survey in a wrinkle in time. At the time of writing, California is experiencing a severe drought and vast fires, which affect growing seasons, mast crops, and animal populations. If we were to sample the area now, we might get different results. This is one reason to save and share results when you have accumulated an extensive database. Chapter 14 tells you various ways you can do it.

HOW LONG WILL IT TAKE TO CAMERA TRAP A PUMA?

New camera trappers are prone to ask this question, and versed camera trappers are prone to offer ballpark answers. If you keep records, you can answer it yourself with little work. I calculated the

Location	CT Days	N of Detections	CT Days/ Detections
Placer Trail	876	10	87.6
The Pass	1188	26	45.6
Cedar Stump	584	20	29.2

Figure 15.10. Puma visits recorded on three trails in the Butte Creek Canyon Watershed. The three cameras operated simultaneously for much of the sampling period and photographed the same animals.

visitation rate for three game trails in a creek catchment near my house (Figure 15.10) and found that I can expect a puma photo every one or two months, depending on location. The three trails differ in slope and direction, but the most frequently used, Cedar Stump Trail, leads to permanent water. You can collect your own data and do this simple calculation, and with a bit more work, you can calculate the probability of detection.

BEARS AS CAMERA PESTS: A BEDTIME STORY FOR INSOMNIACS

Bears are notorious camera pests (see "How Bears Mess with Cameras"). Their engagements with my cameras varied in effort and length, but I viewed them with angst because paws and jaws can lay low almost any camera, and a wet nose can smudge a lens. When I secured my cameras in metal security boxes, I watched new bear videos with guarded confidence. It seems that few bears become bored with cameras. They may not examine every camera they encounter, but in my experience, been there, done that never applies to a bear's camera curiosity.

During the Covid-19 pandemic, I reviewed the footage from eight trail sets and five water sets taken between 2013 and 2020. Figure 15.11 shows the data fields and method of measurement. The

#	Data	Notation
1	Clip ID	e.g., 776.03_06-09
2	Date of Record	e.g., 7/25/2013
3	Camera Make	Bushnell, Browning, Sony s600, GoPro, etc.
4	Type of Set	Trail or Water
5	Artificial Light	1=Yes, 0=No
6	Direction of Movement	Coming or Going
7	Cam-Subject Distance	in Feet
8	POV	Subject or Watcher
9	Age/Sex Class	C1, C2, Adult Male, Adult Female, Sex Unknown
10	Reaction Target	Camera, T-post, LED Array, Bear-guard (Without Camera)
11	Reaction	1=Yes, 0=No
12	Pauses & Looks	1=Yes, 0=No
13	Apps & Sniffs/Flehmen	1=Yes, 0=No
14	Paws	1=Yes, 0=No
15	Bites or Nudges	1=Yes, 0=No
16	Avoids or Runs Away	1=Yes, 0=No
17	Rubs Cam and/or Tree	1=Yes, 0=No
18	Description of Clip	Text

Figure 15.11. Bear-camera spreadsheet data fields were arranged as columns (#). Observations were entered in each row as I viewed the video clips.

first column lists the camera sets and clip numbers and links every observation in the spreadsheet to specific media files. The second column records the date the clips were taken, which allows me to sort data by month or season. Columns 3–8 contain information about camera variables, and columns 9–17 note the age–sex class and behavior of the subjects. I aged bears as cubs and yearlings, and classed bears larger than yearlings as adults, which I sexed when possible. If bears did not react to the camera (data field 11), I left the remaining columns blank.

I was pleased with the volume of information, but the spreadsheet was a head-scratcher. My habit of changing camera sets added new variables and complicated the database. I was looking only for age and sex differences in bear "camera decorum," but the data were not collected under standardized conditions.

HOW BEARS MESS WITH CAMERAS

Bears don't engage every camera they encounter. When a bear walks past a camera, it's hard to tell if it ignored it or failed to detect it. Most bears in my neighborhood glimpse cameras discreetly. Suspicious bears like Shy Boy that eyeball cameras, test the air, and decamp are uncommon.

More common are curious bears that engage cameras to various degrees. A brief sniff at close range may satisfy their interest. They may also sample cameras daintily with the tip of the tongue, which often segues to a slack-jawed expression. These actions probably involve the Jacobson's organ. The heavy breathing and nose-blowing associated with camera examinations presumably liberate scent from the camera's nooks and crannies and invariably overload the microphone.

Chemical investigation can escalate to test the camera's durability. Bear cubs usually have to stand bipedally to paw and scratch the camera, while larger bears flex a foreleg to arm-wrestle. Tenacious adults apply brute strength with both forelimbs, which can snap wooden stakes and bend steel T-posts. The tugging and pushing seems to embrace the mechanical principles of Icelandic wrestling, except that bears can't topple a staked camera by lifting it out of the ground.

Biting bears often target the infrared array and the Fresnel or camera lens. Images of palate, tongue, and teeth are sometimes captured when a bear jaws the whole camera, and sometimes they are good enough to decorate the dentist's waiting room. The nose pad is also an effective prodding tool. Bears use it to poke cameras with enough force to clank a metal security box and can easily upend cameras on skid mounts by pushing with the nose (Figure 15.12).

Sometimes the bear–camera dalliance culminates in a bout of body-rubbing. Usually, they target the camera and its mount with the head and neck, but the back, rump, and sides also come into play. My cameras didn't always capture rubbing in video, but it was easily detected by the rhythmic sound and sway of the clip.

The after-effect of a bear's camera engagement is an object changed by the addition of bear scent, and that may well add to the camera's inherent appeal to other bruins.

Figure 15.12. Glimpses of bears getting to know cameras, as recorded by the victimized cameras. ***A***, who goes there?; ***B***, up close and personal; ***C***, the pushover; ***D***, the big sniffer closes in; ***E***, a beady eye evaluates and takes its measure; ***F***, the paw has its way; ***G***, the breath test and the pearly gates. (***A*** and ***B***: Napa County; ***C–G***: Butte County, California, California; ***A***, ***B***, ***D***, ***E***, and ***F***: Sony s600; ***C*** and ***G***: video screenshots, Browning Recon Force Extreme)

For example, I had set cameras at different distances at water and trail sets. The cameras differed in appearance and in the type of lighting (white or infrared). Camera perspective depended on its distance to the bear, its angle to the animal's path of movement, and whether the bear approached it frontally or from behind. These variables raised more questions than I wanted to answer, but after parsing and sorting data in the spreadsheet, I learned a few things.

LESSON 1: NOT ALL SETS ARE EQUAL

It was plain to see that bears behaved differently at trail and water sets, even though the cameras were mounted similarly—on tree trunks or posts 18 inches to 5 feet above ground and 1–50 feet from trails and pools in creeks. Cameras on or right next to trails were an attraction that bears couldn't overlook, though eating acorns or body-rubbing nearby trees could delay camera encounters.

Cameras near water were a secondary attraction because drinking and bathing were usually the animals' priorities. Pools were often sites of relaxed interludes where bears groomed, snacked on thirsty wasps and bees, played with sticks and stones, and sometimes dozed for more than an hour. They might approach and explore the camera staked near the water's edge before moving on to other locations. Because of these differences, I decided not to combine data from trail and water sets.

LESSON 2: THE RISKS OF ILLUMINATION

Analysis would have been simpler if my cameras were identical. I used commercial trail cameras (Browning and Bushnell) and homebrews (Panasonic and GoPro), which may look as different to bears as they do to people. They may also smell different, but they all stirred reactions, especially at night when the infrared light switched on. A few bears startled and took flight, while confident bears paused before giving cameras a once-over. Lights normally activate in dim lighting and at night, but a few homebrewed cameras triggered white flash and floodlights in daylight. When bears reacted to them, it was immediate.

I hypothesized that cameras draw the attention of bears more often when they emit visible light (infrared or white light) than when they don't. Figure 15.13 compares reactivity to presence or absence of camera lights in the four age–sex classes. Cubs, females, and males reacted more often when the lights were on than off. Fewer observations were made at night because bears are mainly diurnal. A larger sample size may have given a different result for yearlings.

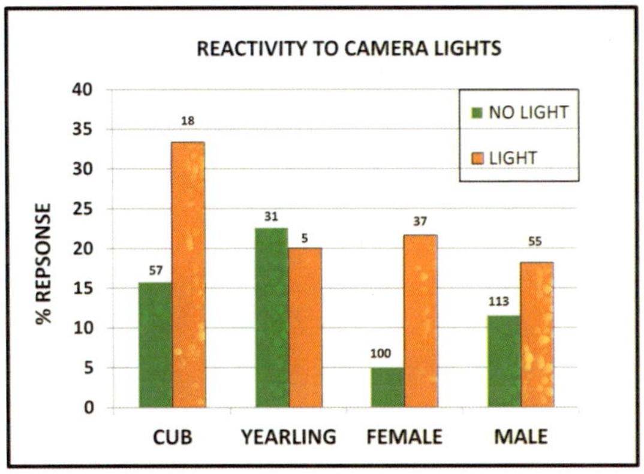

Figure 15.13. Age–sex differences in reactivity of bears to camera lights. Bar height is the percentage of observations when bears reacted to cameras. Values above the bars are the number of observations. Reactions varied from approaching and engaging the camera, to looking at it, and even occasionally fleeing.

LESSON 3: POINT OF VIEW MATTERS

Cameras recorded a first-person **point of view (POV)** when they were both recorders and targets of the bears' attentions.[7] Commercial trail cameras in particular recorded baffling montages of limb and facial anatomy accompanied by loud soundtracks. The POV was personal and dramatic, but behavioral details were hard to discern. The wide-angle lenses of homebrews revealed more, but I couldn't see the whole from the parts until I deployed watcher cameras 14–35 feet away from the subject cameras.

The third-person POV provided by watcher cameras showed whole-body sequences of bears engaging another camera at a distance. They revealed actions like biting, nose-bumping, and nose-blowing that were easier to hear than see in the first-person POV of the subject cameras. Water sets were the only situation in which I had enough videos to compare the watcher and subject cameras.

The watcher cameras recorded higher percentages of behavior than the subject cameras (Figure 15.14). Pawing was the one exception. Avoidance reactions (avoid/bolt) are noteworthy. The watcher cameras recorded avoidance more than twice as often as the subject cameras did. A possible explanation is that spooky bears absconded before they were close enough to trigger the subject cameras. The larger and lateral FOV of the watcher cameras detected animals more effectively than the frontal or head-on perspective of the subject cameras (see chapter 6 for more on motion sensors).

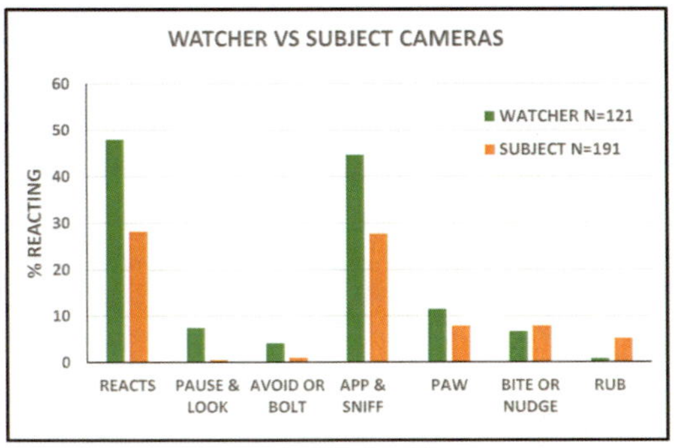

Figure 15.14. Differences in reactivity and specific reactions of adult bears recorded by watcher and subject cameras. Data were collected at water sets. (*APP* = approach; *N* = number of visitations)

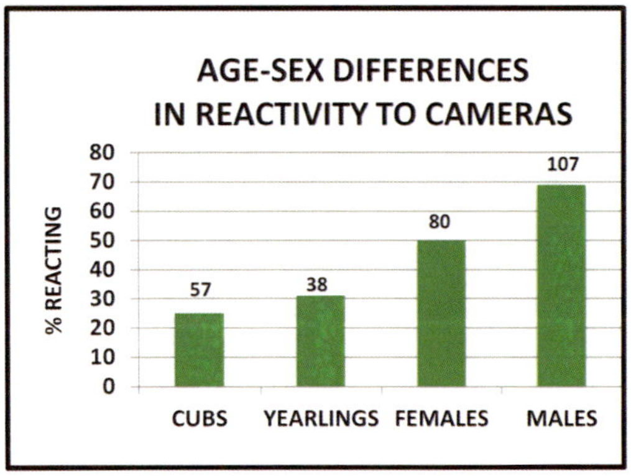

Figure 15.15. Age–sex differences in responsiveness to cameras by bears coming and going on trails. Values above the bars are the number of observations. The height of each bar was calculated by dividing the number of times bears reacted by the number of observations.

HOW OFTEN DO BEARS REACT TO CAMERAS?

I expected young bears to be more curious than adults, and adult females to be more curious than males. At least that was my impression from viewing videos. My measure of curiosity was whether or not they reacted to the camera. I analyzed data from subject cameras that were six feet or less from the trail or water. (Trail data were of bears approaching the front of the camera.)

The results (Figure 15.15) didn't match my expectations. Adults reacted to cameras more often than cubs and yearlings, and males reacted more often than females. Yearlings reacted slightly more often than cubs. I had thought females were more reactive than males. Playing with data makes you question what you have seen and can expose your hidden biases. I now realize I need a more precise measure of curiosity, like duration of camera engagement, to prove myself wrong.

CAN I PROTECT CAMERAS FROM BEARS BY SETTING THEM OFF THE BEATEN PATH?

Data from paired cameras allowed me to compare the percentage of reactions to cameras at different distances to pools on seasonal creeks. I sorted the data from cameras set within 8 feet of the pool and those set 10–26 feet away. The subjects were twenty-nine identifiable animals, and the mean number of observations per animal was thirteen with a range of four to twenty-nine. Figure 15.16 shows that cubs, females, and males reacted more often to cameras nearby. Observations of yearlings were too few to be useful. This was an exercise in proving the obvious, but it showed that despite their relaxed detachment at bath time, adults and cubs are aware and reactive even to distant cameras.

If analyzing data is your goal, design simple experiments. Examining observational data can shed light on a problem, but it may not satisfy your curiosity completely. If nagging questions remain, plan a simple experiment and collect more data. Analysis will be simpler, and you will learn more.

Figure 15.16. Reactions to cameras near (< 8 ft) and far (8–26 ft) by age–sex class. The numbers above the bars represent the percentage of observations.

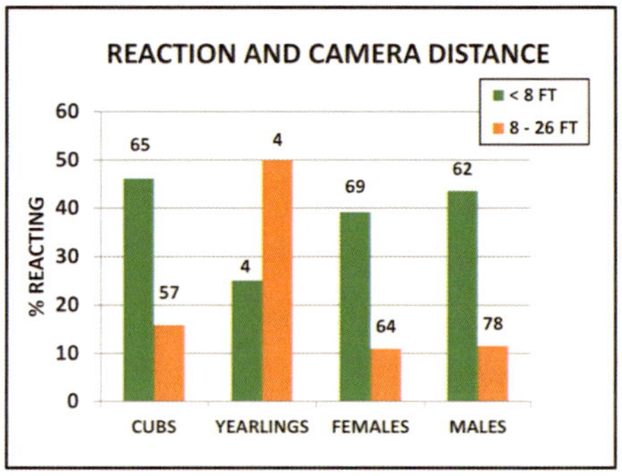

Parting Thoughts

I wrote this book at the pace of a tortoise—slow, vulnerable to distractions, and outpaced by new information. Cameras continued to change, and I had to revise sections and delete links to expired URLs. It seemed a never-ending process. The proliferation of commercial trail cameras is a bewildering gift. We greet the annual arrival of new models like we welcome the migrating birds of spring, but there's a difference. Cameras appear in greater numbers and diversity each year, while the count of migrating birds dwindles. Ecologists like to remind us that "everything is connected to everything." The human footprint grows and nature recedes.

Making the perfect trail camera has been elusive. Companies usher in long-awaited improvements with a certain fanfare and usher out some other features we dearly value. This curious coincidence agitates grumbling in cyberspace. Business responds to squeaky wheels and products improve. Today commercial trail cameras serve most of us better than ever. And for professional wildlife photographers and amateur wannabes—like yours truly—there will always be electronics wizards to turn high-class retail cameras into superb camera traps. I think Shiras, Champion, and Gregory would gladly adopt trail cameras as scouting aids and tools of research, but I suspect they would turn to modern high-end homebrews for the photographic satisfaction they were accustomed to.

Let us visit our three camera trapping pioneers in the twilight zone. I see omniscient gray-haired spirits in a rustic cabin visiting the here and now. They converse politely and marvel at society's many improvements, but they find the state of our planet unsettling. They were believers in progress, but technology and growth have levied a heavy toll. They view humanity now as an ominous horde, and they long for the wild places that are long gone.

Their conversation seeks the comfort zone of common interest, and another paradox pops up: in the good old days, camera traps were alternatives to hunting; modern trail cameras are hunting accessories. This strange new age has much to offer, but the world has changed in so many mysterious ways. The things they valued most, untrammeled nature and wildlife, are fewer now than they were back in the day. Society tries to rescue charismatic species that stand in the way of *progress*, but it fails to grasp the vital connections of ecology. The pace of change is outrageous, and they are glad they were men of the past. I think they would take heart if they could know of the many who follow in their footsteps. We are their kindred spirits quietly seeking magic in the wild places—because for us the world is still a place where there is hope for discovery beyond the far ridges.

Beyond the far ridges, male Sharp-tailed Grouse still compete for hens on traditional breeding grounds. May native prairie and this rite of spring always be there to make us wonder. (West central Saskatchewan, Canada; Canon Sl1 with two external flashes). *Courtesy of Jeremi Skelton*

Acknowledgments

Many people helped me write this book, but I'll start by thanking my daughters for planting the seeds. Jill urged me to "stop emailing photos to friends and just start a blog," and Lauren suggested that I "see if SF State wants a class in camera trapping." My dear wife, Shirlee, the third and central pillar in my support system, never sought publicity but was subjected to sundry cameo appearances in the blog as "the redhead." These three women have given my life a certain balance that would have otherwise escaped me.

Blogging introduced me to people with shared interests and to hunting and photography forums, most notably *Real Deal Hunting Chat*, *Hag's House*, and *Camtrapper*, which focuses largely on wildlife photography. I want to thank Sean Hall and Jeff Dale for piloting *Camtrapper* over the years.

San Francisco State University's Sierra Nevada Field Campus was a wellspring of naturalists and creative earth muffins. I thank its two directors, Jim Steele and his successor J. R. Blair, for supporting the camera trapping workshops (and a subsequent class on Sierra Nevada mammals) for ten years. Reinforcement from friends and colleagues spurred me on, and with each successive class, the scope of the book grew.

In 2013 Jim kindly hosted a reunion of aging SF State biology majors at the Field Campus. This was Rich Tenaza's brainchild, and it evolved into annual festive gatherings that were generously hosted by Terry and Kimberley Schulz and dedicated to our shared histories and camera trapping. Tenaza, Terry Schulz, Reno Taini, and the late Dave Shumaker were the *dramatis personae*, while the more rational voices were those of spouses Pat Shumaker, Kimberley Pitman-Schulz, Randi duBois, and Shirlee Wemmer.

I am also indebted to another group of individuals who contributed to the book's content by facilitating opportunities to camera trap in new locations.

In 2007 Gary Fellers walked me through the application process to camera trap the furtive mountain beaver (*Aplodontia rufa*) at Point Reyes National Seashore and advised me on coexisting with the Tule elk that shared the coastal scrub.

In 2008 Skip Schwartz, then director of Marin County's Audubon Canyon Ranch, permitted Tenaza and me to survey birds and mammals there. Gwen Heistand was our genial liaison. This is where the camera flashed a klatch of lady volunteers on their weekly walk through the redwoods. I hope their surprise encounters with the "camera troll" amused them as much as their reactions entertained me.

That same year Rich Tenaza opened the gate to the Cleary Reserve (Napa County). The ecological setting had matured since our visits as undergrads, and wildlife had sacked and reclaimed the buildings. Colleagues Steve Anderson and Rod Jackson and the "oven-lovin' woodrat" contributed to a timeless ambience.

Brian Miller and family hosted us at the Wind River Ranch (now Mora River Wildlife Refuge) in northeastern New Mexico in 2008 and 2010. I camera-trapped on both occasions, and Dave Rentz kicked off an insect survey on the second trip. Brian guided us to a bear den and a beaver colony and loaned me the faithful canine Luna to spirit me into the rimrock. You can't ask for a better friend. He shipped me my cameras when the batteries burned out and donated twenty Cuddeback trail cameras for surveys in California.

In early 2009 Shirlee and I were guests on Greg McMillan's ranch in Gillis Canyon, San Luis Obispo County. It was an honor to explore the landscape where his father and uncle, both naturalist ranchers, conducted the first California condor survey and where Carl B. Koford conducted research for his classic book.

Later that year I succumbed to Craig Fiehler's entreaties to survey mammals at the Chimineas Ranch in San Luis Obispo County. My reticence had been the sixteen-hour round trip, but that vanished when I saw "the spread," which is managed by California's Department of Fish and Wildlife. Craig and Ken Hickman were boon companions and first-rate botany instructors. Ken kindly enlisted his family to write a friendly program to extract EXIF data from my sets. Bob Stafford's all night kit fox surveys were awesome acts of professional devotion. Despite a speeding ticket, remedial driving school, a flat tire in the boondocks, and the necessity to buy a new car, I have never regretted the Chimineas years.

In 2010 Jack Dumbacher graciously brokered my meeting with the late Lowell Diller to camera trap red tree voles on Green Diamond Resource Company property in Humboldt County. I thank Eric Forsman for advising me where and how to find tree vole nests and can never repay Terry and Kimberley Schulz for lodging and feeding me during the next few years.

In 2011 Steve Cordes kindly permitted Ken Hickman and me to camera trap in the Upper Butte Basin Wildlife Area (Glenn County). It was a wetland adventure that lasted a year.

In early 2012 Collin and Kathi Harris welcomed Bill Wilson, John Warden, and me to camera trap on their ranch in Antelope Valley (Sierra County). They told us the points of interest and turned us loose, and we soon reveled in the willow thickets and the ancient Pacific Juniper groves nearby.

That summer Carl Hansen and Ginger Gier hosted Shirlee and me at their cabin in Whitefish, Montana. Carl kindly tended the cameras into the fall and then packed and sent them home by mail. Two years later we convoyed the Alcan Highway together to Alaska, camera trapping along the way. We couldn't have found more affable travel mates. More recently Carl set up and tended a video camera at my behest through the winter and skied road-killed deer to the site. Friends like that are hard to come by.

I also thank Tim Jensen (Monterey County Regional Parks) for his interest and efforts to rediscover the last wooden grizzly bear trap built in the nineteenth century. Reno Taini and the late Tony Colonnese joined me on visits in 2015 and 2016. Tragically, an errant bulldozer destroyed the historic site during the Soberanes Fire and that foreclosed our plans to camera trap the erstwhile haunts of "California grizz" in Steinbeck country.

In 2018 Eveline Larrucea and Lucy Blake invited a few codgers to camera trap the Lemon Canyon Ranch in Sierra County and the adjacent wildlands of Plumas County. Eveline still informs us of new opportunities, tends cameras in winter, and is a steadfast source of sound advice. Since then Bill Wilson, Carl Butz, and I have joined her at biannual camera trapping "rendezvouses." Ever since my first camera trapping workshop, Bill has been a continuing source of information and companionship. Carl, the distinguished editor of *The Mountain Messenger*, has even chronicled a few of our catches.

In 2020 Andrea Craig gave me the pleasure of assisting The Nature Conservancy in a camera trap survey of Dye Creek Preserve (Tehama County). Project Manager Scott Hardage was instrumental in carrying it out and hosted a trail camera workshop in 2022. That same year the Rhode Island Natural History Survey selected me for its Excellence in Wildlife Conservation, Technology, and Art Award. It was a great and unexpected honor. I am indebted to the Survey's President, Dr. David Greggs, and Program Administrator Kira Stillwell for their personal support and for creatively advancing the study and enjoyment of natural history.

Good neighbors, like-minded spirits, and old friends aided me in other ways. Neighbor Richard Chaddock introduced me to welding and tinkering, made gadgets, and helped in more ways than he knows. For their patience in advising me on sundry electronic challenges, I thank Roger Bailey, Jeff Dale, Ralph Davis, Ron Davis, and Gary Kohout. Dan Holzner opened my eyes to Canon's Hacking Development Kit (CHDK). Jonny Armstrong generously donated equipment. Rodney Jackson, a source of inspiration over the years, loaned me cameras and equipment. For good company on outings and wildlife tips, I thank Phil Howard, Bob Edwards and Bea Heiloway, Ted Trimble and Judy DeNicolo, Dave and Robin Hopkins, Marcel Grondahl, Paul and Helene Baker and their daughter Celeste, Scott Gates and family, Jim Pallette, Peter Hoch, and lifelong friend Dave Rentz.

I have benefitted greatly from comments, editorial suggestions, and text contributions of many colleagues. I thank readers Jonny Armstrong, Peter August, Benjamin B. Beck, Diana Bell, Joel Berger, Gordon Buchanan, Gordon Burghardt, James Champion, Erin Chapelle, Lee Christianson, Jeff Dale, Justine Evans, Henry de Lange, Bill Goodson, Ken Hickman, Paul Joslin, K. Ullas Karanth, Roland Kays, Sebastian Kennerknecht, Andrew Kitchener, Eveline Larrucea, John MacKinnon, the late Chuck McDougal, Woody Meristem, Dietland Mueller-Schwarze, Brian Miller, David Neils, Duncan Parker, Kathy Ralls, the late Galen Rathbun, Cindy Roessler, Francesco Rovero, Terry Schulz, Jeremi Skelton, Dale T. Steele, Rich Tenaza, Andrew Tilker, Duncan Parker, Peter Pearson, Carol Ralph, Francesco Rovero, Peter Van der Naillen, Arthur Veitch, and Neal Wight.

I am indebted to colleagues from the Smithsonian Institution for assistance in locating critical literature, contacts, and wildlife identification: Carl Hansen, Pamela Henson, Margie Gibson, and

Polly Lasker, Don E. Wilson, and Alfred Gardner (USFWS). I am also grateful to the late George B. Rabb (Brookfield Zoo), Amber King and Dawn Roberts (Chicago Academy of Sciences), David Hulse (Ford Foundation), and Inge van Noortwijk (Naturalis Biodiviersity Center, Netherlands). I also thank Mark Rosenthal for diplomacy in facilitating contacts in Chicago.

Other kindred spirits have buoyed my interest with forum posts and emails: Jonny Armstrong, Kalon Baughan, Patrick Burns, Daryl Dean, Edwin Gould, the late Chuck McDougal, David Neils, Casey Smartt, Dawn Tanner, Gary Schroyen, and Cliff Wheeler.

Finally, to the numerous unnamed people who responded to emails and helped me verify the facts, please know that you are appreciated.

I am indebted to BRG Scientific, a publisher and brain trust that lurked and worked to help me get the lead out. I could always rely on my trusted friends Ted Grand for sage advice, Pete August for solutions, and Scott Derrickson for reality checks. Professional editors Stan Carey and Melissa Stevens made each and every chapter better, Carol Underwood kindly crafted the eye anatomy figure on short notice, Shelley Quattrocchi skillfully indexed the finished product, and Becky Bayne worked magic on the layout and graphic design. Without them, *The Companion* would be just another unfinished project.

Notes

Chapter 1

1. Wemmer, "Not-So-Giant Civet," *Camera Trap Codger* (blog), February 19, 2009. An account of our expedition in 1980; Wemmer, "30-Year-Old Possum Picture," *Camera Trap Codger* (blog), February 12, 2009. A description of the camera trap we used in Indonesia.

Chapter 2

1. Shiras, *Hunting Wildlife with Camera and Flashlight*, 1: vii.
2. Shiras, *Hunting Wildlife with Camera and Flashlight*, 1: 29.
3. Shiras, *Hunting Wildlife with Camera and Flashlight*, 1: 33
4. Shiras, *Hunting Wildlife with Camera and Flashlight*, 1: 38.
5. Shiras, *Hunting Wildlife with Camera and Flashlight*, 1: 33.
6. Shiras, *Hunting Wildlife with Camera and Flashlight*, 1: 37.
7. Shiras, *Hunting Wildlife with Camera and Flashlight*, 1: 76.
8. Wikipedia, "Carl Georg Schillings," last updated June 29, 2022; Schillings, *With Flashlight and Rifle*, 1905; Becker, *Carl Georg Schillings, Jäger und Naturschützer*, 26: 93-104.
9. Shiras, *Hunting Wildlife with Camera and Flashlight*, 1: viii.
10. Gregory, *Biographical Memoir of Frank Michler Chapman*, 116–119 and 133; Chapman, "Who Treads Our Trails?" 52(3): 331–345; Chapman, *My Tropical Air Castle*, 196-219.
11. Shiras, *Hunting Wildlife with Camera and Flashlight*, 1936.
12. Gregory, *Biographical Memoir of Frank Michler Chapman*, 130.
13. Shiras, *Hunting Wildlife with Camera and Flashlight*, 2: 257–276.
14. Shiras, *Hunting Wildlife with Camera and Flashlight*, 2: 276.
15. Morrisey, Rory, email to author, December 15, 2022; Morrisey, *Oliver G Pike*, 2.
16. Estate of Oliver G Pike, "Oliver G. Pike," Oliver G Pike (website), accessed December 6, 2022.
17. Bright, *100 Years of Wildlife*, 2007.
18. Champion, *Jungle in Sunlight and Shadow*, 97; Champion, *Tripwire for a Tiger*, 202. A primary source of Champion's biographical information and writings compiled by his grandson, James Champion; Wangchuk, "How a Forest Officer Clicked India's First Photo of a Tiger in the Wild," *The Better India*, June 16, 2022.
19. Champion, *With a Camera in Tiger-Land*, 206.
20. Champion, *With a Camera in Tiger-Land*, 75.
21. Champion, *With a Camera in Tiger-Land*, 212
22. Champion, *With a Camera in Tiger-Land*, 214.
23. Champion, *With a Camera in Tiger-Land*, 210.
24. Champion's grandson James compiled his grandfather's writings in *Tripwire for a Tiger: Selected Works of F. W. Champion*. The book includes a bibliographic sketch of F. W. on pages xiii to xxxiii.
25. Champion, *Jungle in Sunlight and Shadow*, 245.
26. Champion, *With a Camera in Tiger-Land*, 78.
27. Champion, *With a Camera in Tiger-Land*, 77.
28. Champion, *With a Camera in Tiger-Land*, 81.
29. Champion, *With a Camera in Tiger-Land*, 81.
30. Champion, *With a Camera in Tiger-Land*, 117.

31. King, *Gregory Brothers Papers*, March 2017.
32. Gregory, W., *Biographical Memoir of Frank Michler Chapman*, 71.
33. Gregory, W., *Biographical Memoir of Frank Michler Chapman*, 72.
34. Gregory, T., "Black Wolf of the Tensas," 6(3): 36.
35. Gregory, T., "Black Wolf of the Tensas," 6(3): 47.
36. Gregory, T., "Black Wolf of the Tensas," 6(3): 51.
37. Gregory, T., "Black Wolf of the Tensas," 6(3): 51.
38. Gregory, T., "Lion in the Carmens," 1(3–4): 73.
39. Young, *Sketches of American Wildlife*, 143.
40. Gregory, T., "Lion in the Carmens," 1(3–4): 79.
41. Gregory, T., "Lion in the Carmens," 1(3–4): 110.
42. Gregory, T., "Lion in the Carmens," 1(3–4): 77–78.
43. Gregory, T., "Lion in the Carmens," 1(3–4): 118.
44. Gregory, T., "Lion in the Carmens," 1(3–4): 119.
45. Gregory, T., "Lion in the Carmens," 1(3–4): 119.
46. Gregory, T., "Lion in the Carmens," 1(3–4): 120.
47. Now named *Thomomys bottae sturgisi* (a subspecies of Botta's pocket gopher).
48. Gregory, T., *Eyes in the Night*, 83–84.
49. Champion, *Jungle in Sunlight and Shadow*, 247.
50. Dugmore, *Romance of the Beaver*, 1914.
51. Champion, *Jungle in Sunlight and Shadow*, 244.
52. Chapman, "Who Treads Our Trails?," 344.
53. Chapman, "Who Treads Our Trails?," 135.
54. Shiras, *Hunting Wildlife with Camera and Flashlight*, 1: 76.
55. Champion, *Jungle in Sunlight and Shadow*, 245.
56. Champion, "Wildlife in the Indian Forests," 40.

Chapter 3

1. Pearson, "Traffic Survey," 40(2): 169–180.
2. Pearson, "Traffic Survey," 40(2): 170.
3. Pearson, "Traffic Survey," 40(2): 170.
4. Pearson, "Habits of Microtus californicus," 30: 231–249; Pearson, "Habits of Harvest Mice," 41(1): 58–74.
5. Abbott and Coombs, "Photoelectric 35-mm Camera," 45(2): 327–330; Buckner, "Preliminary Trials," 78(2): 77–79; Dodge and Snyder, "Automatic Camera Device," 24: 340–342; Osterberg, "Activity of Small Mammals," 43: 219–229; Ray, *Portable Automatic Camera System*, 59.
6. Wemmer, "Camera Trap Pioneers: Nicholas Smythe," *Camera Trap Codger* (blog), January 31, 2010. Smythe was a fellow grad student at the University of Maryland; quotes are from email correspondence in 2010.
7. Wemmer, "Camera Trap Pioneers: Nicholas Smythe," *Camera Trap Codger* (blog), January 31, 2010.
8. Wemmer, "Camera Trap Pioneers: Nicholas Smythe," *Camera Trap Codger* (blog), January 31, 2010.
9. Smythe, *Natural History of Central American Agouti*, 52.
10. Joslin, Paul, email to the author, April 26, 2007.
11. Joslin, Paul, email to the author, April 26, 2007.
12. Joslin, Paul, email to the author, April 26, 2007.
13. Joslin, Paul, email to the author, April 26, 2007.
14. Joslin, Paul, email to the author, April 26, 2007.
15. Joslin, Paul, email to the author, April 26, 2007.

16. Gross, *Donald R. Griffin*, 20. A summary of Griffin's brilliant career and his quest to understand animal consciousness. The beaver experiment was a footnote in his busy life and is not mentioned in the memoir.
17. Müller-Schwarze, Dietland, email to author, November 3, 2016.
18. Müller-Schwarze, Dietland, email to author, July 25, 2011.
19. Choegyal, "Charles McDougal," *Nepali Times Buzz*, May 13–19, 2016.
20. McDougal, Charles, email to author, December 21, 2009.
21. Smith et al., "Metapopulation Structure of Tigers in Nepal," 176–189.
22. Karanth and Nichols, *Methods for Monitoring Tiger and Prey Populations*, 303; Karanth, "Estimating Tiger Populations," 333–338; Karanth and Nichols, "Estimation of Tiger Densities," 79(8): 2852–2862; Karanth, "Trouble with Tiger Numbers," 55–59.
23. McDougal, Charles, email to author, December 21, 2007.
24. Goodson, W., email to author, May 30, 2017.
25. Goodson, W., personal communication with author, January 18, 2017.
26. My account of camera hacking is based on information gained during my entry into the field of camera trapping in the 1980s and started with hacking a 35mm point-and-shoot camera based on information found on *Jesse's Hunting Page*.
27. These ideas emerged from forums such as *Hag's House* and *Real Deal Hunting Chat* (no longer online) and *Camtrapper.com*. A succession of controllers came on the North American market: Pixcontroller, Bigfoot, YetiCam, Snapshot Sniper, TRLcam.com, and RC Davis Game Camera Solutions. I've benefited greatly from discussions with Ralph Davis, Ron Davis, and Jeff Dale.
28. Kucera and Barrett, "History of Camera Trapping," 9–26.

Chapter 4

1. Galen Rathbun related details about sengi discoveries and taxonomic status in person and by telephone.
2. Stümpke, *Snouters*,1981.
3. Rovero and Rathbun, "Potentially New Giant Sengi," 95(2): 111–115; Rovero et al., "New Species of Giant Sengi 274(2): 126–133.
4. The species has also been dubbed the Vu Quang Ox, which conjures the image of a bulky species of cattle. It's evolutionary position within the family Bovidae remains to be clarified.
5. Hance, "The Saola," April 4, 2011; Vu Van Dung, and Du Tuoc, "Discovery of the Saola," 11–13.
6. Vu Van Dung et al., "New Species," 363: 443–445.
7. Robichaud, "Physical and Behavioral Description," 79(2): 394–405.
8. Tilker, Andrew, email to author, April 13, 2022.
9. Flux, "Sumatran Rabbit *Nesolagus netscheri*," 137–139.
10. Surridge et al., "Striped Rabbits in Southeast Asia," 400: 726.
11. Bell, Diana, email to author, December 17, 2014.
12. Tilker, Andrew, "Fading Stripes," 2018; An Nyuyen et al., "New Records," 85(5): 417-421.
13. Andrew Tilker, email to author, April 13, 2022.
14. McCarthy et al., "Using Camera Trap Photos,"46(3): 438-441.
15. Smithsonian Insider, "New Species of Giant Rat," September 9, 2009.
    Gordon Buchanan told me how he photographed the rat in February 2013 during a BBC film crew shoot in Myanmar.
16. Mirsky, "Kha-nyou smell a rat?" July 25, 2005.
17. Jenkins et al., "Morphological and Molecular Investigations," 2(4): 419–454.
18. Dawson et al., "*Laonastes* and the 'Lazarus effect' in recent mammals," 311: 1456–1458.
19. Huchon et al., "Multiple Molecular Evidences," 104(18): 7495–7499.
20. Stephenson, "Retired FSU Professor," June 6, 2006.
21. Holden, "New Species in Borneo?" 310: 1764; Mariani, "Unexplored Forest Reveals Mystery," December 9, 2005; Fair, "Scientists Foxed by New Carnivore,"24(1): 30.

22. Naish, "Mystery Mammal of Kayan Mentarang," *Scientific American* (blog), February 25, 2013.
23. Meijaard, Kitchener, and Smeenk, "'New Bornean Carnivore,'" 36(4): 318–324.
24. Owen, "Photos: Cameras 'Trap' Hairy-Nosed Otter," July 26, 2010.
25. Coleman, "Debbie Martyr Rediscovers Extinct Muntjac," October 10, 2008.
26. Wildlife Conservation Society, "In Iran, Camera Traps Reveal Rare Asiatic Cheetahs," *ScienceDaily*, September 12, 2005. Unfortunately, a lot has happened since then, and the future of cheetahs in Iran is not particularly hopeful.
27. Dinata et al., "Camera Trapping Rare and Threatened,"18(1): 30–37; Anon, "Cuckoo Caught on Camera," 28(3): 9.
28. Nguyen et al., "Camera-Trap Evidence," 3: 1650–1654.
29. Rovero and Zimmermann, eds. *Camera Trapping for Wildlife Research*, 2016. Graph on page 2.
30. Stager, "Role of Olfaction," 1964.
31. Hall, Sean, email to author, July 16, 2008, concerning three deer and bat photos taken at a mineral lick in Floodwood, Minnesota, "surrounded by a marshy lower area, hence the extensive mosquito population."
32. Wemmer, "Bat Strafes Owl," YouTube, July 23, 2003.
33. Knörnschild and Tschapka, "Predator Mobbing Behaviour," 18(2): 1132–1135.
34. Wemmer, "Wet and Thirsty Raptors," YouTube, September 25, 2013.
35. Hickman, "Coast Packrats," *Nature of a Man* (blog), January 25, 2012.
36. de Lange, Henry, emails to author, April 11, 2011, and March 4, 2017.

Chapter 5

1. *Camera trap* is an older term that is often used generically, while the terms *trail camera*, *scouting camera*, and *game camera* were invented to describe commercial products. Camera traps are remote cameras, but not all remote cameras are camera traps. Many DSLR cameras can be controlled remotely with a handheld transmitter.
2. There are several varieties of trail camera reviews. Amazon's Questions and Answers are helpful, and the comments and ratings of satisfied and disgruntled buyers give further insight. Other reviewers rank cameras on a scale and list the manufacturers' specifications. To my knowledge, the most comprehensive trail camera reviews based on technical testing can be found on TrailcamPro.com.
3. Hunting Farmer, "DIY Solar Trail Camera Power," YouTube, February 20, 2019; Game Cam Wildlife, "How to Add an External Battery," YouTube, February 6, 2021.
4. For a description of trail camera functions and how they are tested visit https://www.trailcampro.com/pages/trail-camera-tests.
5. Traditionally, a photographic flash referred to the brief pulse of light (measured in milliseconds) produced by flash bulbs and electronic strobe lights. Longer duration light has been called continuous light. Traditional photographic floodlights are an example of this.
6. I was unaware of this until I set three Browning Recon Force Extreme cameras side by side on a single mount. When I viewed the footage, I heard the metallic click of neighboring cameras.
7. These functions are called *Dynamic* in Bushnell models and *IR Smart Video* in Browning models.
8. Among point-and-shoot cameras available in 2022, the following models can be hacked as infrared cameras: Sony models P32, P41, W5, W7, W50, W200; Kodak V803 and V1003; Fuji JZ100 and JZ250; and Samsung PL200 and PL210.
9. For information on adjusting sensitivity in inexpensive commercial PIR sensors, see "Testing a PIR" at https://learn.adafruit.com/pir-passive-infrared-proximity-motion-sensor/testing-a-pir. This information is particularly useful to homebrewers.
10. This feature is currently available only in recent Browning models up to a maximum limit of five minutes.
11. Cambridge in Colour (https://www.cambridgeincolour.com/tutorials/image-interpolation.htm) provides a detailed explanation of interpolation in digital photography.
12. The most commonly used microcontrollers are various models of the opensource Arduino controller, available from https://www.adafruit.com/arduino and https://www.sparkfun.com/ as well as other independent dealers on eBay and Amazon.

13. Dale, "TRLcam's Economy DSLR Camera Trap," *Camtrapper* (blog), March 12, 2012. The first of a five-part series on making a relatively inexpensive DSLR camera trap; Dale, "DSLR and Mirrorless Camera List," *Camtrapper* (blog), April 4, 2014. Jeff Dale's list of DSLR cameras suitable for use as camera traps, with comments by members of the *Camtrapper.com* forum; Burrard-Lucas, "Turn your DSLR Camera," *Digital Photography School* (website), accessed December 6, 2022. An introduction to using DSLRs as camera traps; the author manufactures accessory equipment; Rondeau, "Professional Camera Trapping," Emmanuel Rondeau.com, accessed December 6, 2022. Website and ordering form for the author's eighty-three-page e-book; Rydell et al, "Bat Selfies," 102(3): 793-809. An excellent reference on photographing bats in flight.

Chapter 6

1. Stockman, *How to Shoot Video*, 2011. This book offers helpful advice for amateur videographers on planning and shooting films; Wildeye School of Wildlife Filmmaking (http://www.wildeye.co.uk/) is an informative website that provides general advice and classes in wildlife and conservation filmmaking. They also sell several books on the subject.
2. Wemmer, "Bears vs. the Squeaker," *Camera Trap Codger* (blog), July 20, 2013. Video showing the failure of a homemade noisemaker ("the squeaker") to alarm curious black bears.
3. A brief summary of how to navigate on foot in the wilderness is available at https://howtowilderness.com/navigation/; Seidman and Cleveland, *Essential Wilderness Navigator*, 2001. A well-written reference on the basics of wilderness navigation.
4. Gregory, "Black Wolf of the Tensas," 6(3): 35–68.
5. Herper is to herpetologist what birder is to birdwatcher. Herpetologists study reptiles and amphibians.
6. More than a dozen books are in print treating North American animal tracks and sign. A few titles are included here. Full bibliographic information is listed on the Recommended Reading List at the back of the book.

    Mammal Tracks and Sign: A Guide to North American Species is the most detailed modern treatment of North American species, illustrated with fine color photos.

    Peterson Field Guide to Animal Tracks is the updated version of the old standby, with pen and ink sketches of prints and scat and some color photos.

    Animal Tracks: A Folding Pocket Guide to the Tracks and Signs of Familiar North American Species (A Pocket Naturalist Guide) is a pamphlet of common mammal tracks using pen and ink sketches with paintings of each species.

    Camera Trapping Guide, Tracks, Sign, and Behavior of Eastern Wildlife is a guide specifically for camera trappers and features photos of tracks and other sign, and tips on camera trapping each of thirty-seven species.
7. Lowrey, "Conservation Dogs Sniff Out Wildlife," *American Kennel Club*, March 3, 2020. An article about the roles that sniffer dogs play in wildlife conservation and disease detection.
8. State and regional field guides of plants provide good descriptions of habitats. State field guides to mammals usually treat habitat as an introductory topic. Older editions often discuss the subject in sections titled life zones, biotic areas, and floral provinces. More recent books treat habitat in sections such as physiographic provinces, climate, and vegetation. These guides also describe habitat in the accounts of species.
9. Hickman, "Pygmies of the Great Basin," *Nature of a Man* (blog), July 24, 2014. Post about camera trapping pygmy rabbits; Rohde, "*Brachylagus idahoensis* pygmy rabbit," 2006.

    Summary of pygmy rabbit biology; Larrucea and Brussard, "Habitat Selection," 89(3): 691–699. On the habitat characteristics of the pygmy rabbit.
10. Nat Geo Live, "Steve Winter: On the Trail of the Tiger," YouTube, March 14, 2012. Inspirational lecture by award-winning photographer Steve Winter of the National Geographic Society on tiger conservation, with anecdotes about his use of DSLR camera traps.
11. Hobby, The Strobist (website), accessed December 6, 2022. While this site is currently in archive mode, it is still a wealth of encyclopedic references on the use of off-camera flash and free online lessons on lighting; McNally, *Hot Shoe Diaries*, 2009. Comprehensive coverage of off-camera flash photography; Winters, *Tutorial in High-Speed Photography*, 1992–1996. An excellent PDF manual for those wishing to explore the use of Vivitar 283 flash units.

12. Bright Hub, "What Is a Slave Flash," May 17, 2010. Somewhat dated, but a good summary of remote optically triggered flash features; Jack the Hat Photographic, "Off-Camera Flash Techniques," YouTube, September 27, 2012.

13. Garrett, "Vivitar 2800 Hack," YouTube, June 27, 2012. Charles Garrett's eighteen-minute tutorial on converting a Vivitar 2800 flash to a slave flash.

14. Smartt, "Bunker Hill Wireless Build," *Camtrapper* (blog), March 16, 2016. Instructions and photos on building a wireless PIR sensor for a homebrewed DSLR camera; Smartt, "Wireless Infrared Beam Break Units," *Camtrapper* (blog), January 20, 2018. Casey Smartt's illustrated post about building a wireless beam break motion sensor for use with Nikon and possibly Canon DSLRs, with URL links and discussion by forum members; Smartt, "New AIR Build," *Camtrapper* (blog), February 11, 2018. Casey Smartt's follow-up post about building a passive infrared sensor for use with DSLR cameras, with remarks and discussion by forum members.

15. Schroyen, "Sony 3N Video Build with LED Lighting," *Camtrapper* (blog), September 8, 2016. Gary Schroyen's (a.k.a. Westcocanuck2) forum post about using motion detector security lights for shooting night video using a Sony 3N video build.

Chapter 7

1. Wildfelids, "Mountain Lion Making a Territorial Scrape," YouTube, July 20, 2016. This video clip shows the tangential view of a mountain lion making a scrape; Parliament of Owls, "Daytime Cats!" YouTube, April 29, 2014. This video clip shows a lateral view of a mountain lion making a scrape.

2. Olszewski, "Role of Uprooted Trees," 19: 99–104. On the differing use of logs as runways by different species of European rodents.

3. Wemmer, "Spooky Cameras and Squirrelly Neophobia," *Camera Trap Codger* (blog), January 8, 2012. Blog post and movie about the unexpected results of a camera trap set on a split-rail fence.

4. Hickman, "Rocky Relationships," *Nature of a Man* (blog), October 22, 2015. This blog post and included photos feature visitors to a pika's haystack in the eastern Sierra Nevada; Wemmer, "Restless Haystack," *Camera Trap Codger* (blog), October 23, 2011. Blog post and YouTube time-lapse video about mountain beaver haystacks.

5. Stone Cabin Photos, "Bears and Gray Fox," YouTube, August 28, 2015. Trail camera video showing these two species attacking a wasp nest at night; ZovTV, "Bears vs. Bees," YouTube, April 29, 2009. Short Russian video of a bear cub attacking an artificial beehive and climbing a tree with the comb; Boot, "Bears vs. Bees—Nature's Secret Power," YouTube, August 17, 2012. A ground nest of wild bees attack and defeat a bear seeking a taste of honey.

6. Lezappen, "Sped Up Timelapse Footage," YouTube, February 26, 2017. Three-minute video of hornet queen building a paper-mâché nest; Peetre, "Paper Wasps Making a Nest," YouTube, July 10, 2012. Three-minute time-lapse of wasp egg-laying taken at 1 frame/4 seconds over a two-day period.

7. Digital Naturalism Laboratories, "Mysterious Castles of Clay," YouTube, November 24, 2013. A survival presentation by Joan and Alan Root.

8. Wemmer, "Aerial Dogfights of Spring," *Camera Trap Codger* (blog), March 24, 2007. Some hints on photographing hummingbirds with a homebrewed camera trap.

9. Wemmer, "Bears on the Puma Kill," *Camera Trap Codger* (blog), December 13, 2017. Blog post and video of bears and smaller scavengers feeding on a black-tailed deer killed by a puma.

10. Viral Hogs, "Red Foxes enjoy Elk Carcass," YouTube, March 3, 2017. Brief, exciting footage of red foxes at an elk carcass photographed with a GoPro left inside the ribcage.

11. State of California, "Rodenticides: Rodenticides Can Harm Wildlife," California Department of Fish and Wildlife, accessed December 1, 2022. Brief discussion of the problem and some frequently asked questions.

12. Root, *Ivory, Apes and Peacocks,* 126. Root describes using inflatable decoys for famous German zoo director Bernhard Grzimek.

13. Story Trender, "Animals in Mirrors Hilarious Reactions," YouTube, June 1, 2015. Amusing clips of African wildlife reacting to their own mirror images; Svendsen and Armitage, "Mirror-Image Stimulation," 54(3): 623–627. Research finding that wild yellow-bellied marmots "reacted intensively with their image" while held in captivity.

14. Grigonis, "'Spy in the Wild,'" 2017. BBC trailer showing the reactions of a troop of Indian langur monkeys to a robotic infant monkey they apparently thought had died; John Downer Productions, "Robotic Spy Puppy

Meets Wild Dogs," YouTube, May 22, 2017. Excerpt from a John Downer production showing African wild dogs reacting to an animatronic pup.

15. Palminteri, "Camera Trapping in the Trees," July 4, 2017. Techniques and experiences of biologists while camera trapping in tropical forest canopies of Latin America; Hart, T., "Eight Months in the Forest Canopy," *Searching for Bonobo in Congo* (blog), September 19, 2017. Blog post about camera trapping monkeys in trees in the Congo.

16. Meiselman and Doyle, "Habitat and Microhabitat," 135: 33–42.

17. Jobanek, "Searching for the Tree Vole," 89(4): 369–400. Interesting history of early attempts to collect specimens of a specialized arboreal rodent.

18. Cornell Lab of Ornithology, "Platform for Success," NestWatch, accessed December 1, 2022. Advice from Cornell University's Laboratory for Ornithology on making an artificial osprey nest; Bohm, "Artificial Nest Platforms for Raptors," 11(4): 97–99. Article on the use of artificial nests by great horned owls.

19. Muir, "Windstorm in the Forests," 1894. For those who enjoy wild landscapes, Muir's moving prose on climbing a tree in a storm is a lesson on risk and inspiration.

20. Raphael and White, "Use of Snags," 86: 1–66; Bull, Roberts, and Torgerson, "Trees and Logs,"1997.

21. The wildlife section of the Connecticut Department of Energy and Environmental Protection (https://portal.ct.gov/DEEP/Wildlife/Fact-Sheets/Nest-Boxes-and-Structures-for-Wildlife) contains many details about artificial nest structures for various species of birds and a few mammals. The Texas Backyard Wildlife YouTube channel (https://www.youtube.com/@TexasBackyardWildlife) has excellent videos of suburban wildlife in nest boxes and constructed dens.

22. Cornell Ornithology Lab, "All Cams," All About Birds, accessed December 1, 2022. To view various species of nesting birds watch the live feeds from the Cornell Laboratory of Ornithology's nest-monitoring cameras.

23. Peeters, *Field Guide to Owls*, 2007. Information about owl biology and nesting habits; Craighead and Craighead, *Hawks, Owls, and Wildlife*, 1969. The old standby reference on raptor biology and nesting habits.

24. Wemmer, "Submariner's Solution," *Camera Trap Codger* (blog), May 12, 2007. Blog post about using a DIY periscope as an aid for aiming a camera trap at an owl's nest hole.

25. Young and Dobyns, *Den Hunting*, 132: 1–8.

26. Wemmer, "Denning Kit Foxes," *Camera Trap Codger* (blog), August 18, 2013. Blog post and video about a kit fox den with pups in San Luis Obispo County, California.

27. North Carolina Museum of Natural Sciences, "Coyote Finds Den of Red Fox Pups," YouTube, July 1, 2015. Excellent presentation by Roland Kays and Stephanie Schuttler on using trail cameras to explore behavior at a natal den of a red fox; French, "Photographer Captures Fight," *Billings Gazette*, February 8, 2020. A red fox boldly defends its den against a badger intent on eating fox pups.

28. Lonie, Gordon, email to author, February 22, 2012.

29. British wildlife camerawoman Justine Evans refreshed my memory of Chris Packham's mole filming accomplishment during a BBC shoot in Burma (February 2013). Packham also narrated a three-part BBC series, *The Burrowers: Animals Underground*, in which replica burrows were created to record secretive behavior. Links to all three episodes are provided in the Bibliography.

30. Easterson, "Burrow-Cams," Video Data Bank (website), accessed December 1, 2022. "Sam Easterson is a video artist who has specialized in capturing animal perspectives using videography." Video Data Bank has a variety of sample clips, including those at this URL of a porcupine and a badger in their burrows.

31. Wemmer, "Video Vision in the Tunnel—Part 2," *Camera Trap Codger* (blog), November 19, 2012. Blog post about making a subterranean video camera set for mountain beaver, with a video showing the results.

32. Wemmer, "Creature Comfort," YouTube, August 8, 2013. Video of a bear sleeping fitfully in a creek on a hot summer day.

33. Struhsaker, "Behavior of Elk," 24: 80–114. An excellent reference on elk behavior, with observations on behavior around wallows.

34. Prior, "Best Moose Documentary Ever," YouTube, March 6, 2017. Russian production about farmed moose—part 4 of the film shows male and female behavior at urine pits.

35. Dale, Jeff, email to author, May 6, 2017; Dale, "All in a Days Work," *Camtrapper* (blog), April 22, 2016. Jeff Dale's thread on *Camtrapper* describing his underwater camera build; Tye et al., "One House," 185(2): 229–240. Contains a description of a solar powered IP camera modified by Jeff Dale for time-lapse data collection of a beaver lodge and the results of the study.

Chapter 7 Sidebar Notes

Scent-Marking

Cassidy and Gonzales, "DNA Testing in Animal Forensics," 69(4): 1454–1462. Discusses the use of non-human DNA to identify animal species for law enforcement.

Thomsen and Willerslev, "Environmental DNA," 183: 4–18. Research on using DNA in environmental samples of soil, sediment, or water (eDNA) to identify the presence of species in the absence of any recognizable remains.

Lasley and Kirkpatrick, "Monitoring Ovarian Function," 22(1): 23–31. Research on monitoring female reproductive cycles using hormonal metabolites in urine and feces.

Quest for the Red Tree Vole

Forsman, Swingle, and Hatch, "Behavior of Red Tree Voles," 83(3): 262–272. The first behavioral study of red tree voles based on video from arboreal cameras.

Wemmer, "Hard-to-Get Tree Vole," *Camera Trap Codger* (blog), January 15, 2012. Blog post about making an arboreal camera set for the red tree vole.

The Lure of Rimrock and Bluffs

This experience is based on visits to the former Wind River Ranch (now the Mora Wildlife Refuge, USFS) located near Watrous, New Mexico, the Chimineas Ranch (managed by California's Department of Fish & Wildlife), and Dye Creek Preserve (The Nature Conservancy).

The Camera Trappin' Blues

The woes of camera trappers are certainly worthy of artistic expression.

Ninja Sets

Personal experiences prove that tempting locations sometimes compel camera trappers to jury-rig camera sets when their equipment is unequal to the task.

Chapter 8

1. Hunters and trappers use the words *lure*, *decoy*, and *bait* for a variety of commercial products. I defined them according to dictionary entries but added biological criteria to distinguish between them. The information is based on numerous threads in camera trapping forums, published references, and personal experience.
2. Farmers Insurance, "Close Claws," Aban Commercials (website), accessed February 7, 2023. This Farmers Insurance advertisement titled "Close Claws" showed the mock use of a laser pointer to distract a mountain lion. The YouTube version, no longer available for viewing, lacked the television disclaimer warning gullible people that a laser could indeed deter a lion attack.
3. Addley, "BBC Series Uses Robot Creature," *The Guardian*, December 30, 2013.
4. Hickman, "I'm on a Perch," *Nature of a Man* (blog), August 5, 2012.
5. Darwin, *Expression of the Emotions*, 1965. Describes orangutans reacting to a mirror at London Zoo.
6. Lehner, "Coyote Vocalizations," 26: 712–722.
7. Alcorn, "On the Decoying of Coyotes," 27(2): 122–126.
8. Maser, *Mammals of the Pacific Northwest*, 1998. The author describes shrews' reliance on touch, hearing, and scent in finding and attacking prey.
9. Biologists now advise birders to learn the songs rather than use playbacks. Birders still imitate the soft alarm note that is shared by many songbirds (*pishing*), as another way to get a better view.
10. Hardy and Morrison, "Factors Affecting the Detection, 28: 333–342; Morrell, Yahner, and Harkness, "Factors Affecting Detection," 19: 481–488.
11. ROD917 [pseud.], "Electronic Predator Call," AutoDesk Instructables (website), accessed December 2, 2022. These plans make use of some commercial electronic products; Cobb, "Homemade E-Caller," *Predator Masters* (forum), accessed December 6, 2022. Robert Cobb's plans for building a predator caller using RadioShack parts.
12. News Corp Australia Network, "Beaver Anal Secretions," September 16, 2013.
13. Farrell, "You Might Have Consumed," April 25, 2020.

14. Trapper supply catalogues on the internet are an interesting vernacular source of information about commercial scent lures and baits.

15. Haagen-Smit, "Essential Oils," 3(1): 71–83.

16. Vedantu, "Jacobson's Organ," Vedantu.com (website), accessed December 2, 2022; Estes, "Role of the Vomeronasal Organ," 36(3): 315-341.

17. Rosell and Sanda, "Potential Risks of Olfactory Signaling," 17(6): 897–904.

18. McDaniel et al., "Efficacy of Lures," 28(1): 119–123. This paper describes the use of carpet patches as scent dispersers and examines the reactions of Canadian lynx and other species to various commercially available scents.

19. Thorn et al., "Estimating Brown Hyaena Occupancy," 39(1): 1-10.

20. Ackerman, *The Bird Way*, 89–109.

21. Turner, "How Does Catnip Work," *Scientific American*, May 29, 2007; Hill et al., "Species-Characteristic Responses," 2(2): 239–253.

22. Reynolds, "Studies of Reproduction," 52(3): 223–284. Description of the opossum's saliva bathing behavior. I have seen this behavior myself, but my opossum also licked and saturated its sides with saliva.

23. Savage, "Science of Scent," *Trailie Paws For Thought* (blog), September 19, 2022. Well-researched article on scent detection in dogs; Roberts, "Sniffer Dogs," *Dog Learner* (blog) July 24, 2020. Covers breed differences in scent detection and various aspects of sniffer dog training; Randler et al., "Baiting/Luring Improves Detection," 10: 2178–2186.

24. Bossert and Wilson, "Analysis of Olfactory Communication," 5: 443–469. This paper gives mathematical explanations about the transmission of scent under different conditions of air movement and introduces concepts that are now used by sniffer dog handlers; Rattlerjen [pseud.], "How Scent and Airflow Works," *Hound and the Found* (blog) February 22, 2012. A helpful illustrated guide to scent movement.

Chapter 9

1. Cambush, "Do It Yourself Trail Camera Mount," *Cambush* (blog), October 12, 2013. Post on making a do-it-yourself trail camera mount.

2. Ken Hickman, creator of the *Nature of a Man* blog (https://natureofaman.blogspot.com/), made use of the "pipe-through camera case" to mount a homebrewed camera on metal stakes using a threaded rod. This design could probably be modified for use with commercial trail cameras using the cable lock channel on the back of the camera.

3. Smartt, "Down-n-Dirty Camera Skid," *Camtrapper* (blog), September 7, 2014. Casey Smartt's DIY instructions for building a low-profile mount for DSLR camera traps.

4. A variety of products can be found on the internet by searching for *telescoping flagpole*.

5. Foxlights (https://www.foxlights.com/) are a commercial flashing-light predator deterrent used by the livestock industry in Australia and beyond. They have also been effective in deterring snow leopard predation on corralled sheep and goats in Asia. To see Foxlights in action visit: https://snowleopardconservancy.org/foxlights-video/.

Chapter 10

1. Accounts of forgetful camera trappers are based on posts on Camtrapper.com.

2. Water Science School, "Snowmelt Runoff and the Water Cycle," USGS.gov, June 12, 2019; Project Wet Foundation, "Discover Yellowstone," USGS.gov, 2012. PDF that explains factors causing floods; Nelson, "Flooding Hazards," last updated October 19, 2015. Technical source of information on flooding hazards, prediction, and human intervention; Californian Nevada River Forecast Center, "Observed Precipitation," NOAA (website), accessed December 2, 2022. An example of a forecast database used to track flood water.

3. Veitch, Arthur, email to author, October 2, 2015.

4. Skelton, Jeremi, email to author, December 13, 2014, regarding coyote chewed cameras; Skelton, "Not a Good Feeling," *Camtrapper* (blog), December 13, 2014.

5. Wight, Neil, email to author, October 10, 2015.

6. Meristem, Woody, email to author, September 26, 2015.

7. Miller, Brian, email to author, October 17, 2015.

8. Brown, "Mountain Lion Steals Game Caller," YouTube, November 12, 2013. Hand-held video of a mountain lion absconding with a mechanical predator lure and caller. Scroll to minute 18.00 to see the hunter finding the caller.

9. Personal observation of footage showing elephants interacting with trail camera during filming of BBC's Wild Burma, January 2013.

10. Hickman, "Stephen Colbert's Nightmare," *Nature of a Man* (blog), October 29, 2011. Bear and gray fox insults.

11. Kingtrout435 [pseud.], "Trail Camera-Ants," Bowsite.com (forum), June 4, 2008. Various posts—and amusing comments—regarding insect infestations in trail cameras, including termites in a DSLR and ants and yellow jackets in trail cameras. Mentions circuit boards as ant attractants; Pjdoc1 [pseud.], "Ants vs. Trail Cams," Hunting Indiana (forum), June 19, 2013. Tree honeydew as an ant attractant.

12. Pearson, "CDC Warning on Misuse of Pest Strips," Wired.com, January 29, 2014.

13. Cremer, "Termites Build a Nest in My Camera," *Camtrapper* (blog), September 4, 2015.

14. Tenaza, Richard, email to author, September 28, 2022.

15. Tenaza, Richard, email to author, September 28, 2022.

16. Wemmer, "On Camouflaged Trail Cams," *Camera Trap Codger* (blog), January 4, 2011. Blog post on shortcomings of commercial trail camera camouflage and on camouflaging a camera with natural materials.

17. Davexx1 [pseud.], "Camo/Hide Your Trail Camera . . .," Georgia Outdoor News (forum), July 31, 2007. Part of a long thread on hiding trail cameras, with ideas including using a hornet's nest as a trail camera disguise and using hidden trail cameras for human surveillance; Cornpatcher [pseud.], "What are your most creative ways," Archery Talk (forum), June 28, 2012. Two redundant forum threads about hiding trail cameras, including the hornet nest disguise.

18. Bigboyz [pseud.], "Trail Camera Anti-Theft," HuntingNet.com (forum), June 23, 2005. Part of a long thread on leaving a photo and a message in the camera that identifies the owner.

19. Staff Writer, "Stolen Trail Camera," *Lincoln Times-News* (website), January 9, 2015; Hinton, "Stolen Trail Camera Leads," KFVS (website), May 12, 2018.

20. Wikipedia, "Eye-Fi," Wikimedia Foundation, last updated January 21, 2023.

21. Stolen Camera Finder, "How It Works," StolenCameraFinder.com (website), accessed December 5, 2022.

Chapter 11

1. Mills, Z., "Ethics of Wildlife Photography," *The Wildlife Collective*, February 2, 2022; Ebersole, "Ethics: Too Close for Comfort," *Audubon Magazine*, May–June 2015.

2. Tinbergen, *Curious Naturalists*, 1968.

3. Palmer, *Shooting in the Wild*, 197.

4. NANPA Ethics Committee, "Ethics," North American Nature Photography Association (website), accessed December 2, 2022.

5. The Nature Group, *Nature Photographers' Code of Practice*, 2007.

6. ASCPF, "International Federation of Wildlife and Nature Photography," ASCPF (website), accessed December 5, 2022.

7. Banisar and Davies, "Privacy and Human Rights," Global Internet Liberty Campaign (website), accessed December 5, 2022. The quoted definition from Alan Westin's book *Privacy and Freedom* (1967) is one of several given in this document.

8. Klowsowski and Fishbein, "Know Your Rights," *Lifehacker*, September 25, 2019.

9. Benson, "Paparazzi in the Woods," *Slate*, August 14, 2008. Quotations on camera traps and privacy.

10. Swedish Environmental Protection Agency, "Game Camera for Private Use," Naturvårdsverket [Swedish Environmental Protection Agency] (website), accessed April 6, 2023.

11. Kelly, "Austrian Politician Photographed Having Sex," *Huffington Post*, June 16, 2012.

12. Anon, "Beim Tresorknacken im Wald gefilmt," [Filmed Cracking the Safe in the Woods], *Burgenland News*, March 20, 2008.

13. Amrhein, "Animal Rights vs. Human Rights," Daily Reckoning (website), accessed December 5, 2022.

14. Sample, "Wildlife Documentaries Infringe," *The Guardian*, April 29, 2010.

15. Gannon, "No Pictures, Please!" Yahoo News (website), October 23, 2013.

16. Verderber, "Camera Trapping: How to Get the Shot," 34.

17. Reed and Merenlender, "Quiet, Nonconsumptive Recreation," 1(3): 146–154.

18. Steidl and Anthony, "Responses of Bald Eagles," 482–491; Stalmaster and Kaiser, "Effects of Recreational Activity," 6(2): 1–46.

19. *The Baobab: Portrait of a Tree* (1973) and *Mysterious Castles of Clay* (1978) by Alan and Joan Root used white lights to capture images of nocturnal galagos, termites, and aardvarks. Dereck and Beverly Joubert's film *Eternal Enemies: Lions and Hyenas* (1992) captured a lion and spotted hyena conflict at night, illuminated with flood light.

20. Wegge, Pokheral, and Jnalwali, "Effects of Trapping Effort," 7(3): 251-256.

21. Yang, "What Causes the Red Eye Effect?" *Yale Scientific*, May 12, 2011.

22. Newbold and King, "Can a Predator See," 36(4): 309–318.

23. Carwardine, "Ethics in Wildlife Photography," *Discover Wildlife*, November 7, 2011.

24. Forsman, Meslow, and Wight, "Distribution and Biology," 87: 3–64; Guynup and Ruggia, "Owls Face Spotted Future," *National Geographic Channel*, July 22, 2004.

25. Bart, "Impact of Human Visitation," 16: 187–192; Clark and Wobeser, "Making Sense of Scents,"28(1): 31–37.

26. Caro, *Antipredator Defenses*, 2009.

27. Lehmberg, "Fox and a Badger Fight in Yellowstone," YouTube, February 24, 2015. Dramatic interaction of a red fox attacking a badger in defense of its burrow and its pups within.

28. Nichols, "Full Disclosure," *Photocrati*, January 30, 2010.

29. Booth, "Wildlife Photographer of the Year," *The Guardian*, January 20, 2010.

30. Palmer, *Shooting in the Wild*, 2010.

31. Williams, "Phony Wildlife," *Audubon Magazine*, March–April 2010.

32. Gasset, *Meditations on Hunting*, 1995.

33. Hall, "What's the Future of Trail Cameras," The Meateater (website), January 20, 2022; Kayser, "Outlawing Trail Cameras," *Outdoor Life*, April 14, 2010.

34. Leopold, *Sand County Almanac*, 1949.

Chapter 11 Sidebar Notes

The Risks of Disturbance

Berger, Daneke, Johnson, and Berwick, "Pronghorn Foraging Economy," 25(3): 193–208.

Frid, "Vigilance by Female Dall's Sheep," 53(4): 799–808.

The Visual Effects of Electronic Flash

Guyton, *Textbook of Medical Physiology*, 1961.

Opstad, "Can Flash Photography Damage Eyes? [Explained It]," NESOP (website), accessed December 7, 2022.

Romer, *Vertebrate Body*, 1962.

Denning Wolves and Sleeping Bears

Argue, Mills, and Patterson, "Behavioural Response of Eastern Wolves," 86(5): 400–406.

Frame, Cluff, and Hik, "Response of Wolves," 71(2): 316–320.

Habib and Kumar, "Den Shifting by Wolves," 272(3): 259–265.

Linnell et al., "How Vulnerable Are Denning Bears," 28(2): 400–413.

Nonaka, *Response of Breeding Wolves*, 2011.

Thiel, Merrill, and Mech, "Tolerance by Denning Wolves," 112(2): 340–342.

Exceptions to the Rule

In 2003 G. Fellers posted a hysterical video titled "Secret lives of wildlife exposed: Mountain lion, long-tailed weasel, and mountain beaver video" on the USGS website. Sadly, the video is no longer available for viewing.

Gackstetter, *Trail Camera Tails of Lincoln County, Minnesota*, 2012; personal communication with the author.

Hickman, Ken, email to author, May 3, 2011.

Texas Penal Code, Chapter 9. Justification Excluding Criminal Responsibility, Subchapter D, Protection of Property, Sec. 9.42. Deadly Force to Protect Property.

Chapter 12

1. Texas Penal Code, Chapter 9. Justification Excluding Criminal Responsibility, Subchapter D, Protection of Property, Sec. 9.42. Deadly Force to Protect Property.
2. Forestry Service, "Know Before You Go," USDA (website), accessed December 6, 2022. Contains USFS guidelines on wildlife photography, but it makes no mention of trail cameras.
3. National Park Service, "Filming and Still Photography Permits," NPS (website), accessed December 6, 2022. Concerns general permits to film in national parks; National Park Service, "Scientific Research and Collections Permits," NPS (website), accessed December 6, 2022. Concerns scientific permits for research, such as wildlife surveys using trail cameras in national parks.
4. US Fish and Wildlife Service, "Laws, Agreements, and Treaties," FWS (website), accessed December 6, 2022.
5. US Fish and Wildlife Service, "Species," FWS (website), accessed December 6, 2022. This site makes it possible to search for endangered species listed by the US Fish and Wildlife Service.
6. Ballotpedia, "Endangered Species by State," Ballotpedia (website), accessed December 6, 2022. This site lists threatened and endangered species by state.
7. Montana Fish, Wildlife, and Parks, *Deer, Elk, and Antelope Hunting Regulations*, 23.
8. Section 86, California Fish and Game Code.
9. California Fish and Game Commission Mammal Hunting Regulations, 2020–2021, §251.1 Harassment of Animals, §251.3 Prohibition Against Feeding Big Game Animals, and §257.5 Prohibition Against Taking Resident Game Birds and Mammals by the Aid of Bait, https://fgc.ca.gov/Regulations/Current/Mammals#251_1.
10. Black, *Black's Law Dictionary*, 2nd ed., 644.
11. Section 257.5, California Code of Regulations, Title 14, Division 1, Fish & Game Commission, Department of Fish & Game, Chapter 1, Prohibition Against Taking Resident Game Birds and Mammals by the Aid of Bait.
12. Section 257.5, California Code of Regulations, Title 14, Natural Resources, Division 1, Fish and Game Commission, Department of Fish & Game, Subdivision 2, Chapter 1, Prohibition against feeding big game mammals.
13. Section 251.3, California Code of Regulations, Title 14, Natural Resources, Division 1, Fish and Game Commission, Department of Fish & Game.
14. Section 2000, California Fish and Game Code, Division 3, Chapter 1, Taking and possessing in general.
15. Section 473, California Fish and Game Code, Title 14, Division 1, Subdivision 2, Chapter 6, Possession of Nongame.
16. Section 3012, California Fish and Game Code, Division 4, Part 1, Chapter 1, Article 1, Methods of Taking.
17. Section 475, California Fish and Game Code, Title 14, Division 1, Subdivision 2, Chapter 6, Methods of Take for Nongame.

Chapter 13

1. Bitterman, "Comparative Analysis of Learning," 188(4189): 699–709. A technical review of learning theory comparing the views of comparative psychologists and ethologists; Casey, "'Learning Theory' Terminology," *Raining Cats and Dogs* (blog), October 12, 2013. Excellent readable account for pet owners on the terminology of operant conditioning, with references to Pavlov and other early animal psychologists; Domjan, *Principles of Learning and Behavior*, 1998. A college textbook on learning concepts and animal behavior.
2. Heinrich, *Mind of the Raven*, 2000. Reference to the author's observation that ravens show fear to events that violate expectation.
3. Curio, "Adaptive Significance of Avian Mobbing," 48(2): 175–183. An examination of the functions of mobbing of predators by birds, especially with regard to effects on the behavior of birds and the predators.
4. Taursir [pseud.], "Squirrel Daylight Robbery I," YouTube, March 15, 2007. Delightful series of videos about English bird lovers defending bird feeders against tree squirrels addicted to bird seed.
5. Viitala et al., "Attraction of Kestrels," 373: 423–425. The first of a series of papers by Finnish biologists on the ability of some predatory birds to detect rodents by UV reflectance of urine and feces.

6. Nams, "Density-Dependent Predation," 110: 440–448. Finding cited in text that captive striped skunks formed search images of food based on scent and that searching time improved with experience.
7. Kloppers, St. Clair, and Hurd, "Predator-Resembling Aversive Conditioning," 10(1): 442–450. The report cited in text on aversive conditioning of elk in Banff National Park.
8. Andelt, "Behavioral Ecology of Coyotes," 94: 1–45. Observation cited in text interpreted as stimulus generalization of human scent resulting in avoidance of scent stations used for surveys.
9. Freeland and Janzen, "Strategies in Herbivory by Mammals," 108(961): 269–289. Important early reference on the evolutionary arms race between herbivores and plants mentioned in text.
10. Gabriel and Golightly, "Aversive Conditioning of Steller's Jays," 79(5): 894–903. Feasibility study cited in text about aversive conditioning as a tool to reduce egg predation of Marbled Murrelets.
11. Runte, *Yosemite: The Embattled Wilderness*, 1990. This book is my main source of information about the history of bear–human conflict in Yosemite; McCullough, "Behavior, Bears, and Humans," 10(1): 27–33. A thoughtful discussion about applying the principles of learning to human–bear interactions in national parks.
12. Breck, Lance, and Seher, "Selective Foraging," 90(5): 1041–1044. The study of black bear break-ins in automobiles in Yosemite National Park.
13. Schneier, "Human/Bear Security Trade-Off," *Schneier on Security* (blog), August 18, 2006. Reference to park ranger quote in text about the relative intelligence of bears and people.
14. Personal observations of hazing and nocturnal ranger patrols of campgrounds are based on twelve years of annual visits to Yosemite.
15. Mazur and Seher, "Socially Learned Foraging Behaviour," 75(4): 1503–1508. A study in two national parks showing that rearing conditions of bear cubs around campgrounds or in the wild strongly influence subsequent choice of foraging areas after independence; Mazur, "Does Aversive Conditioning Reduce," 74(1): 48–54. National Park Service research on the effectiveness of different aversive conditioning methods on black bears in Sequoia National Park.

Chapter 13 Sidebar Notes

Coyotes Bold and Shy

Dobie, *Voice of the Coyote*, 147. Natural history observations of coyotes and people in the Southwest.

Harris and Knowlton, "Differential Responses of Coyotes," 79(11): 2005–2013. Wild coyotes are more likely to respond to artificial scent stations on the boundary or outside of their home range than within their home range.

Mettler and Shivik," Dominance and Neophobia," 102(1–2): 85–94.

Murie, *Ecology of the Coyote*, 1940. Comments on effects of hunting and protection on tameness and wildness in coyotes.

Sequin et al., "Wariness of Coyotes," 81(12): 2015–2025. The behavior of radio-collared coyotes of differing social status and age in response to the presence of camera traps.

Windberg, "Coyote Responses," 74(12): 2248–2253. Captive coyotes are more neophobic to visual cues in familiar than unfamiliar areas.

Lessons Learned by the Loner Wolves

Young, *Last of the Loners*, 1970. Young was a federal predator control agent who chronicled the demise of the last loner wolves.

Gipson, Ballard, and Nowak, "Famous North American Wolves," 26(4): 808–816. A critical examination of the loner-wolf phenomenon and a debunking of previous claims of loner longevity.

Chapter 14

1. Clapham et al., "Automated Facial Recognition," 10(23): 12883–12892; Kays et al., "Which Mammals," 103(4): 767-775.
2. Ritchie, "How to Upload Way Points," YouTube, September 7, 2016. This video demonstrates how to upload from your GPS receiver.
3. Cadle, "Upload GPS Coordinates," YouTube, January 25, 2016. This video demonstrates the use of CVS data table to upload a series of waypoints, icons, and descriptions.
4. Hindy, "10 Best Excel Apps," Android Authority (website), April 29, 2022.

5. Thomson, Potgieter, and Bahaa-el-din, "Closing the Gap," 56(4): 721–739.
6. Agarwal, "Tools for Managing EXIF Data," *Digital Inspiration* (blog), May 20, 2012. An article explaining EXIF data; Anand, "24 Best Free EXIF," List of Freeware (website), accessed December 6, 2022. Description of free EXIF viewer software for Windows.
7. Graphic Converter (website), accessed December 6, 2022. Information about a versatile EXIF reading program for Mac and Windows computers.
8. Creative Commons, "About," Creative Commons (website), accessed December 6, 2022.
9. International Press Telecommunications Council, "What Is Photo Metadata?" IPTC (website), accessed December 6, 2022.
10. Smoller, "Using Orphan Work," *Artrepreneur* (blog), March 11, 2021. General information about orphan works.
11. Quinn, "Sample DMCA Take Down Letter," *IPWatchdog* (blog), July 6, 2009. This site summarizes how to locate webhosting companies and compose a takedown letter to curtail infringement of copyright.
12. Hart, "History and Philosophy of Project Gutenberg," Project Gutenberg (website), accessed December 6, 2022.
13. iNaturalist, "Help," iNaturalist (website), accessed December 6, 2022. Detailed answers to frequently asked questions about the organization.
14. American Society of Mammalogists, "Mammal Images Library," American Society of Mammalogists (website), accessed December 6, 2022.
15. US Fish and Wildlife, "Volunteer Opportunities," FWS (website), accessed December 6, 2022; National Park Service, "Citizen Science," NPS (website), accessed December 6, 2022.

Chapter 15

1. Kucera and Barrett, "History of Camera Trapping," 9–26.
2. Rovero and Zimmermann, eds, *Camera Trapping for Wildlife Research*, 2016. Researchers use different intervals to define detection events, which is discussed in detail in chapter 5.
3. Verts, *Biology of the Striped Skunk*, 1967. Thanks to Lee Christianson for bringing this observation to my attention.
4. Randler et al., "Baiting/Luring Improves Detection," 10(11): 2178; Schlexer, "Attracting Animals to Detection Devices," 263–292.
5. Colleagues and I compiled the number of days that precede the first appearance of four common predators on the Chimineas Ranch. We made 133 camera sets in all months of the year at widely spaced locations using bait or scent lures as attractants. Cameras operated for 4,383 camera days.
6. Tobler et al., "Evaluation of Camera Traps," 11(3): 169–178. Discusses calculation of capture probability.
7. Point of view (POV) refers to a literary voice of narration, e.g., first, second, or third person, or a visual perspective recorded on camera. Filmmakers simulate different perspectives through choice of lens, camera position, and juxtaposition of an image of reference. A wildlife film might create a squirrel's POV by showing in full frame an alarmed squirrel barking high on a limb, followed by the squirrel's arboreal view of a prowling predator; See also Grand, *Carving Nature at Its Joints*, 87. Grand discusses perspectives on page 87 and illustrates their differences in Figures 6.1 and 6.2.

Glossary

A

active infrared sensor (AIR): A sensor that transmits an infrared beam to a receiver and triggers a camera or event recorder when the beam is broken.

active space: A volume of air containing scent molecules in a concentration detectable by a particular animal. An active space in a breeze forms a cone shaped volume, which dog trainers call a scent cone.

adaptive lighting (also automatic dimming): A feature of commercial trail camera video that optimizes exposure of an animal at different distances from the camera.

alkaline battery: The widely available batteries used for electronic devices and named for the alkaline electrolyte potassium hydroxide. The most common sizes are designated as AAA, AA, C, and D cells.

altricial: Born or hatched in a helpless state of development and highly dependent on parental care. See *precocial*.

ambient light detector: A sensor that responds to light and triggers an electrical circuit when illumination falls below a certain threshold. Also called a *day-night sensor*.

aposematism (aposematic coloration): Warning coloration; the distinctive coloration (e.g., skunks, coral snakes, hornets) that warns predators of danger after one or more unpleasant encounters.

Arduino: An Italian-made open-source, programmable controller designed for hobbyists.

associative learning: A learning principle in which an animal recognizes the connection between two stimuli. Classical conditioning is an example.

B

bait: An edible substance used to attract wildlife.

black flash: See *no-glow*.

blank image: Any camera trap image lacking an animal subject. Blank images are triggered by moving air, by animals that left the FOV before the picture was taken, or by animals that were detected outside of the FOV. See *false triggers*.

Blitzpulver: Flash powder, invented in Germany, and the basis of the original flash used with cameras.

blur reduction: A trail camera mode that reduces blur in moving animals. High video speed (60 fps) and fast motion settings also reduce blur. Single frames from videos taken in these settings show animal images with less "stretch."

burst mode: A trail camera shooting mode that fires a rapid volley of images with intervals of less than a second. Also called *multi-shot mode* or *rapid fire mode.*

C

Camera Axe: An open-source hardware and software tool for photographers seeking high-speed images. Camera Axe uses various kinds of sensors to trigger the camera.

camera set: The placement of the camera trap with respect to its field of view, the position of natural props, and the anticipated movement and position of animals.

castoreum (castor): The gummy, pleasantly aromatic secretion of the beaver's castor gland.

Cat 5 or Cat 6 cable: Ethernet cables designed for connecting devices to networks and transmitting signals for telephones and video. Cat 5 cables support transmission speeds up to 100 MHz; Cat 6 cables support up to 250 MHz.

CCTV (closed-circuit television): Video cameras that transmit signals to a limited number of monitors, usually for surveillance of human activity. CCTV is also suitable for remote wildlife videography.

CHDK (Canon Hacking Development Kit): An open-source firmware project that allows users of Canon point-and-shoot cameras to access advanced functions found in professional models of Canon cameras.

chevrotain: Several small species of antlerless deer (family Tragulidae) and close to the evolutionary stem of antlered deer (Cervidae); native to South and Southeast Asia and central and East Africa. Also called *mouse deer.*

classical conditioning: A kind of operant conditioning in which an animal forms an association with a natural stimulus like food with an unrelated stimulus like a sound. Repeated pairing of the two stimuli is necessary for conditioning to take place.

cognition: High-level brain processes involved in perception, learning, and problem-solving.

commensal(ism): A kind of relationship between species characterized by coexistence and benefit to one species (the commensal) while causing no harm to the other.

commercial trail camera: A ready to use camera produced by a commercial company.

conditioned food (taste) aversion: A form of learning where illness caused by eating a particular food generates a strong avoidance of that food. A food poisoning effect and a management tool used to reduce predation on livestock.

conspecific: A member of the same species.

control group: A group of subjects in an experiment that is not subjected to the independent variable tested in the treatment groups.

controller: A general term for an integrated circuit board.

convection: The movement of groups of molecules in fluids and air, or the transfer of heat by wind and currents.

corvid: Members of the crow and jay family (Corvidae), including ravens and magpies.

crane mount: A specialized camera mount for arboreal sets with an extendable boom that can be lengthened and raised.

critical habitat: Habitat within the geographic range of a species that is vital for its conservation. Critical habitats are an important factor in the recovery of endangered species.

D

data strip (or stamp): The information strip that records moon phase, temperature, barometric pressure, and GPS coordinates on the image or video clip; a menu option on most trail cameras.

day-night sensor: See *ambient light detector.*

decoy: A live animal or its likeness that attracts animals, usually predators. As a verb, it refers to the act of attracting wildlife.

descriptive statistics: Statistics used to summarize a collection of measurements broken down into central tendency (mean or average, median, and mode) and dispersion (range or minimum and maximum measures, variance, standard deviation, etc.)

detection angle: The horizontal angle of the detection zone, which is determined by the features of the Fresnel lens, specifically the Fresnel zones and the curvature.

detection event: The criterion used to define independent visits of photo-captured subjects to a camera set. Biologists typically use one hour or longer as the criterion to separate a succession of photos into more or less independent events for statistical analysis. Multiple photos of a species of interest are defined as a single event if they meet the chosen interval criterion. Thus, a succession of photos separated by less than an hour is deemed as a single detection event.

detection zone (field of detection): The three-dimensional space in which a sensor is able to detect a signal.

discrimination learning: Learning to distinguish between similar stimuli.

DSLR (digital single lens reflex): Digital cameras that employ a mirror and prism to allow viewing the image directly through the camera lens.

duiker: Small antelopes (family Bovidae) of the African forest and bush that resemble Asian muntjacs in ecology.

E

ecotone: The border or transition zone between two or more habitats. Also called *edge habitat.*

Endangered Species Act of 1973 (ESA): Federal legislation protecting fish, wildlife, and plants listed as endangered or protected based on technical review. The ESA requires species recovery guided by written plans and mandates cooperation from state and governmental agencies.

Energizer Ultimate Lithium (lithium) AA cells: Nonrechargeable but very long-lasting AA cells that perform well in sub-freezing temperatures.

essential oils: Essence-emitting or scented oils extracted from plant tissues or animal secretions.

EXIF (exchangeable image file format data): The industry standard for formatting images from digital devices. EXIF data are embedded in JPEG images and include date and time, camera make and model, settings, and a thumbnail image.

expansion post: A DIY metal post made from several connectable or telescoping pipes that can be expanded with a screw to fit tightly between rock surfaces. The camera is mounted on the post.

F

f-stop: The focal length of the lens divided by the diameter of the aperture. The size of the lens aperture determines depth of field. The amount of light reaching the focal plane is inversely related to f-stop. Incremental steps in f-stop are equal to a halving or doubling of illumination.

false trigger: A vernacular term for camera trap images that are blank and thus lack animals. *Blank image* is a more accurate term because PIR sensors respond to heat radiating from animal bodies and inanimate objects, and from heat being carried in the air.

field of view (FOV): The scene revealed in the camera's viewing screen or LCD.

field scan: Another term for the shooting mode of trail cameras known as time-lapse.

firmware: Small software programs contained permanently in an electronic device, such as a camera or a camera trap controller, that control its basic operation. Firmware programs are built into a microcomputer chip and can be rewritten and upgraded.

fixed-focus lens: A lens with high depth of field that cannot be automatically or manually focused. Focus is fixed at the hyperfocal distance, with images being acceptably clear from infinity to half the hyperfocal distance.

flash light: A dated term that was used for the explosive illumination created by magnesium flash powder or the device used to produce it. The term was used before the invention of flash bulbs and electronic flash.

flash range: The maximum distance of effective illumination produced by an electronic or LED flash.

Flehmen response: The German word for the lip-curled facial expression made by various species of mammals while sampling urine and other substances with the Jacobson's organ.

flight distance: The usual distance between a predator and prey that elicits the prey's flight or escape.

Fresnel lens: The plastic lens that focuses moving heat on the PIR sensor. Simple Fresnel lenses have a single set of concentric rings (the Fresnel zone), while wide-angle lenses have several Fresnel zones and a convex surface.

G

gimbal: A gyroscope that allows a moving camera to remain horizontal despite angular movements of the mount.

H

habitat generalists/specialists: Species that are able to thrive in many habitats or only a few habitats. Many species also exist in an intermediate number of habitats.

habituation: A form of learning in which an animal's response to a particular stimulus wanes with repeated exposure.

hack: To modify hardware or software; to open a camera and modify circuits for use as a camera trap.

herper: A person who is interested in or studies herpetology (reptiles and amphibians).

Hickman mount: A pipe-through case attached to a threaded rod mounted to a metal concrete stake that allows the LCD to be viewed by opening the back of the camera.

high definition (HD): A video display resolution measuring 1280 x 720 pixels or greater. Smaller displays are called SD (standard definition); 4K or ultrahigh resolution refers to 3840 x 2160 pixels. See *resolution*.

homebrew(ed) camera: A vernacular term for a camera hacked and constructed from electronic components for use as a camera trap.

hot shoe: A camera fixture for the attachment and triggering of an electronic flash unit.

hypothesis: The statement of an idea that can be tested by experimentation. A null hypothesis predicts no difference in the outcome of different treatments. An alternative hypothesis predicts an expected difference in the outcome of the treatments.

I

image sensor: Electronic devices that convert patterns of incidental light into electronic signals representing an optical image.

incandescent: White light produced by a heated metal filament.

infrared (IR): Nonvisible thermal radiation or light having a wavelength greater than visible light (700–1050 nanometers).

inquiline: Animals that cohabit in the nests and dens of other species.

interpolation: A microprocessor function of many trail cameras that automatically enlarges the image and fills the gaps between the original pixels with new color-matched pixels. It's a digital sleight of hand that increases the number of pixels but does not improve photo resolution.

IP (internet protocol) camera: Digital video cameras that produce high-definition imagery by transmitting data from the image sensor through an ethernet cable to a router.

IR cut switch: The infrared filter in digital cameras necessary to render daylight color accurately. In trail cameras the filter is mechanically removed at night in order to detect infrared radiation. Also called an *IR exchanger*.

IR glare: The obtrusive glare most commonly seen in still pictures and videos when the camera calculates exposure for the background and overexposes small objects close to the camera.

IR Smart Video: A Browning trail camera setting that overrides the selected video length and continues to record up to five minutes if the trail camera detects continuing PIR events.

J

Jacobson's organ: A small sensory receptor in the roof of the mouth also known as the *vomeronasal organ*. In mammals it is used to sense nonvolatile compounds and is often associated with a characteristic facial expression. See *Flehmen response*.

JPEG: The most common compression format of digital images. The JPEG file format compresses images for storage and restores them with minimal perceptible loss of image detail.

L

latency to first detection: The elapsed time, usually in days, between setting the camera and the first photo-capture or detection of a particular species.

laser periscope: A homemade gadget used for aiming a pole mounted camera at an elevated nest cavity in a tree. The periscope, made from a short length of one-inch square tubing and a mirror, is attached to the camera with Velcro. A laser beam from ground level is reflected at a right angle through the periscope at the tree's trunk and the pole's height and angle is adjusted.

LCD (liquid crystal display): The electronic flat panel display of digital cameras that allows viewing images and menus.

LED (light-emitting diode): An electronic component that produces light when a voltage is applied to it. Trail cameras contain arrays of infrared LEDs for photography in low lighting conditions.

low-self-discharge (LSD) battery: A battery having a small spontaneous rate of discharge, like the Sanyo Eneloop. It can retain 85 percent of its power after a year on the shelf.

live prey survey: Biologists use live domestic mice as bait to photograph and confirm identification of owls at close range. They attract the owl to the area by imitating the call or by playing the call on a sound recorder or cell phone.

low-glow: The trail camera industry uses this term for infrared lights with wavelengths that are less than 940 nanometers. Low-glow LEDs emit dim red light that is barely visible to humans and probably many mammals. See *no-glow*.

lure: A sight, sound, or scent cue that attracts wildlife. In this book, lures are distinguished from baits, which are foods and thus elicit a feeding response.

M

mast: A generic term for the fruit of trees, shrubs, and vines. *Hard mast* refers to fruit with hard seed coats (acorns, nuts, pinecones), and *soft mast* refers to fruit with fleshy seed coats (berries, cherries, apples).

median: The middle value of a dataset, where half the values are equally divided above and below it.

motion blur: Blurred images that result from low shutter speed. Motion blur often affects only a portion of the image, such as the head or leg of a walking animal. Also called *image stretch*.

motion detector: Any number of sensors that respond to the motion of animals, either by responding to infrared energy, radar, or by breaking a photoelectric or laser beam.

muntjac: A small- to medium-sized solitary forest deer (family Cervidae) widely distributed through South and Southeast Asia.

Myotis: The genus of mouse-eared bats (family Vespertilionidae), the largest and most widespread group of insectivorous bats occurring in the Old and New World.

N

neophobia: Literally "fear of the new." The fearful or distrustful reaction of animals to strange objects or situations, often exhibited as avoidance.

niche: The ecological role of a species within a community.

NiMH (nickel-metal hydride) battery: The most commonly used rechargeable battery cell, which is two to three times more powerful than a NiCd (nickel cadmium) cell. The discharge rate of NiNH cells is greater than that of lithium ion ("Li Ion") cells, but the eneloop cell developed by Sanyo is a low-discharge version of the NiMH cell.

no-glow: Trail camera lights with infrared wavelengths greater than 940 nanometers are invisible to people and probably many mammals and nocturnal birds. Some no-glow arrays emit a deep purple light visible to people only at close range, but these wavelengths can be clearly seen when viewed with a digital camera. Also called *black flash*.

O

one-trial learning: The uncommon situation where an animal forms a strong and lasting association after a single and powerful stimulus-response experience. Also called *single-trial learning*.

open-source: Noncommercial sharing of hardware and software resulting from participatory development.

operant conditioning: A general term for learned responses resulting from punishment or rewards (i.e., reinforcement). Specific behaviors are either strengthened or weakened by rewards or punishment.

P

paca: A large, short-tailed, nocturnal species of rodent (*Cuniculus paca*, family Cuniculidae) found from southern Mexico to southern Brazil.

parallax: The spatial difference in two lines of sight, like the lens and the viewfinder on a twin lens reflex camera. When using a laser periscope to sight a camera on a pole, it is the difference between the position of the laser beam and the center of the photographic target, such as a nest hole. Parallax is minimal when the periscope is next to the camera lens.

passive infrared (PIR) sensor: An electronic detector that responds to the motion of infrared radiation usually after passing through a Fresnel lens.

pattern recognition: The acquired ability of predators to recognize unique visual properties of cryptic or distasteful and toxic prey species. Examples include birds that learn to prey on moths with camouflaged coloration or learn to avoid preying on distinctively marked distasteful butterflies.

PICAXE: A commercially available programmable microcontroller system used by electronic hobbyists.

picture (or photo) delay: See *time-out.*

pipe-through: A security feature of homebrewed camera cases where a half-inch pipe placed through the case allows a cable attachment to a tree and prevents opening the camera case.

pixel change detection: A feature of Canon cameras and the CHDK in which the shutter is triggered in response to changes in the activation of pixels within a sensor grid. Also a feature of IP cameras.

point-and-shoot: Small automatic digital cameras designed for ease of operation.

point of view (POV): The perspective captured by a photo based on the camera's position in relation to the subject—bird's eye view (aerial), up close and personal (macro), eye level, or mouse's view (looking up).

Polaroid camera: A group of American made "instant cameras" that produces self-developing prints shortly after the picture is taken.

precocial: Born or hatched in an advanced state of development and requiring less parental care than altricial species. In the spectrum of condition at birth, many species fall between precocial and altricial.

printed circuit board: A laminated board of copper foil and nonconductive epoxy that uses embedded conductive tracks to connect electronic components such as resisters and transistors.

pugmark: The footprint of a large cat (e.g., tiger or leopard).

pyroelectric: A material that generates a temporary voltage when heated or cooled. PIR sensors contain pyroelectric materials that generate voltage in response to impinging infrared radiation or heat.

R

radio frequency (RF): The electromagnetic frequency band used to transmit radio signals (10 kHz to 300 GHz).

RAM (random-access memory): A computer's temporary data storage, or buffer space, where cameras store images and videos before saving them to the SD card or internal memory.

RAM mount: Commercial brand of ball-and-socket mounts manufactured for cameras and other equipment.

rapid fire: An alternative term for the trail camera setting called *burst mode*.

ratel: The honey badger (*Mellivora capensis,* family Mustelidae) found in Africa and less commonly in South Asia.

RAW: A digital image format that retains most of the dynamic range of the image and thus resembles the film negative.

reaction (or response) time: The interval from detection of a PIR event to shutter release and exposure of the image. Also called *trigger speed*.

ready mode: A camera state in which all circuits are ready to shutter an image or shoot a video. Also called *awake mode*. See *sleep mode*.

rechargeable (batteries): Batteries that are chemically designed to be electrically reversible with the use of an AC charger. External batteries of the same voltage also can be used to recharge batteries in a camera trap.

rechargeable lithium ion (li-ion) batteries: The rechargeable battery used for many electronic devices, known for high capacity, low self-discharge rates, and no memory effects. Many are designed for specific products. The 18650 battery is an all-purpose battery shaped like, but larger than, AA cells.

red-eye: The ghoulish red reflection of human eyes recorded in photos taken without pre-exposure flashes. Cameras with red-eye setting largely eliminate this effect.

reinforcement: A psychological term for reward or punishment that affects an animal's behavior in a predictable way.

remote flash: A secondary external electronic flash triggered by a light sensor when the camera's (master) flash fires. Remote flashes, formerly referred to as slave flashes, generate extra lighting at night and often highlight shape and contour better than the camera's internal flash.

resolution: The number of pixels in a digital photographic frame measured in width x height (e.g., 1280 x 720). See *HD*.

rhodopsin: An enzyme produced in the retina of the vertebrate eye that helps the eye see in low-light conditions.

S

Schmidt Detective camera: An American, medium-to-large format, wooden box camera patented in the 1880s that was used by some of the pioneering camera trappers.

search image: The visual, scent, and sound cues that predators use when learning to locate their prey. Search images result from discrimination learning.

sengi: A group of small to cat-sized African mammals also known as elephant shrews (family Macroscelididae), distinguished by a long mobile snout and insectivorous and often territorial habits.

sensor: Electrical devices that detect heat, light, sound, or other physical phenomena and convert the information into an electronic signal for measurement or control of a device. The sensitivity of a sensor is its ability to detect very fine differences in the measured quantity.

shutter speed: The duration of a camera's image exposure, or the time during which the shutter is open.

sky island: An isolated mountain range over 3,000 feet in elevation inhabited by plants and animals different from those of the surrounding lowlands and often characteristic of higher latitudes.

sleep mode: A power saving function that activates automatically when a camera or electronic device is idle. Sleep mode deactivates the LCD and other operations until the user activates the camera. See *standby mode*.

snag: A dead or dying tree; an important resource for birds and mammals that den or nest in cavities.

snorkel: A plastic or metal tube used in homebrewed cameras to accommodate lens movement made necessary by the trend of extendable zoom lenses on digital point-and-shoot cameras.

solenoid: An electromagnetic switch activated by a current; solenoids were formally the means of depressing the shutter button on film cameras modified as camera traps.

standard deviation: One of several statistical measures of dispersion of values around the mean. Graphed values look like a bell-shaped curve if the data are normally distributed. A low standard deviation means that most data points are close to the mean and that variation is low.

stealth camera: Models of trail cameras that feature low-glow or no-glow infrared arrays for nighttime photos.

steel rebar pin (also concrete stake): A round steel stake measuring 3/4-inch in diameter and 18 –36 inches in length with predrilled holes designed for staking concrete foundation boards. Also suitable with minor modification for staking small camera traps.

stimulus generalization: A kind of learning in which animals conditioned to react to one stimulus react in the same way to other similar but unconditioned stimuli.

strobe light: A general term for a camera's electronic flash, usually a xenon flash lamp, that discharges light from a capacitor and lasts only a few milliseconds.

T

take: A concept introduced in the Endangered Species Act and broadly defined as an action that kills, harms, or injures wildlife. It also includes indirect actions that affect resources essential for a species' survival, such as water, food, and shelter.

tapetum lucidum: The light-reflecting layer of the vertebrate eye that enhances night vision and reflects the light from electronic flashes as eyeshine; it has been likened to automobile headlights on high beam. The hue of the reflected light varies slightly between species.

Tiger Tops: A privately owned jungle lodge in Chitwan National Park, Nepal, specializing in nature tourism.

time-lapse: A method of slow motion cinematography in which naturally slow events are photographed at intervals equal to or greater than one fps. When played at normal rates of 24–30 fps, slow processes like blooming flowers are accelerated. Also the name of a free image-analysis program for field biologists who use camera traps for research.

time-out: A camera trap delay setting between still photos or video clips during which time the camera does not respond to detection events. Also known as *picture delay* or *photo delay*.

tonquin oil: A sweet smelling aromatic extract from the seeds of a Brazilian leguminous tree (*Dipteryx odorata*) used as a food flavoring and a wildlife attractant.

trap-happy: An animal that repeatedly enters traps to get the bait, or a predator that habitually visits traps looking for captured animals.

trigger speed: See *reaction time*.

t-test: A simple statistical test that determines whether the means of two independent populations are significantly different. The variable being tested should be normally distributed. Also called the *Student's t-test*.

TTL (through-the-lens metering): The electronic light sensor in modern cameras that can automatically function as a light meter and allows several modes of metering electronic flash.

U

USB connector: The variety of commonly used and industry standard plugs and ports used to connect computers to electronic devices, such as digital cameras.

V

variable: A factor or feature that can be controlled, measured, and analyzed in an experiment. Scientists study the reaction of their subjects to independent variables by assigning them to different treatment groups.

W

walk test: A feature of camera trap controllers that allows the user to determine the width and depth of the PIR sensor's field of detection. When the user walks within the detection zone, the controller signals PIR events with a flashing LED. Also called *motion test*.

watcher camera: A camera trap set to observe the reactions of animals to another camera.

white light: Visible colorless light having a wavelength of 400–700 nanometers (all wavelengths of the visible spectrum).

windfall: An abundant accumulation of food, such as a fallen fruit, a beehive, or a die-off of migrating salmon.

Bibliography

Books, Journals, and Periodicals

Abbott, Herschel G., and Allyn W. Coombs. "A Photoelectric 35-mm Camera Device for Recording Animal Behavior." *Journal of Mammalogy*, 45, no. 2 (May 20, 1964): 327–330. https://doi.org/10.2307/1377020.

Ackerman, Jennifer. *The Bird Way: A New Look at How Birds Talk, Work, Play, Parent, and Think*. New York: Penguin Press, 2020.

Addley, Esther. "BBC Series Uses Robot Creatures to Document Secret Lives of Animals." *The Guardian*. December 30, 2013. https://www.theguardian.com/media/2016/dec/31/bbc-robot-creatures-spy-secret-lives-animals-wildlife-series.

Alcorn, J. R. "On the Decoying of Coyotes." *Journal of Mammalogy*, 27, no. 2 (May 1946): 122–126. https://doi.org/10.2307/1375230.

Allen, Maximilian L., Micaela S. Gunther, and Christopher C. Wilmers. "The Scent of Your Enemy Is My Friend? The Acquisition of Large Carnivore Scent by a Smaller Carnivore." *Journal of Ethology*, 35 (January 2017): 13–19. https://doi.org/10.1007/s10164-016-0492-6.

Andelt, William F. "Behavioral Ecology of Coyotes in South Texas." *Wildlife Monographs*, 94, (October 1985): 1–45. https://www.jstor.org/stable/3830638.

Anon. "Cuckoo Caught on Camera." *World Birdwatch*, 28, no. 3 (2006): 9.

Argue, A. M., K. J. Mills, and B. R. Patterson. "Behavioural Response of Eastern Wolves (*Canis lycaon*) to Disturbance at Homesites and Its Effects on Pup Survival." *Canadian Journal of Zoology*, 86, no. 5 (May 2008): 400–406. https://doi.org/10.1139/Z08-013.

Bart, J. "Impact of Human Visitation on Avian Nesting Success." *Living Bird*, 16 (1977): 187–192.

Becker, M. "Carl Georg Schillings: Jäger und Naturschützer [Hunter and Conservationist]." *Bongo Berlin*, 26 (1995): 93–104.

Berger Joel, Dennis Daneke, Jim Johnson, and Stephen H. Berwick. "Pronghorn Foraging Economy and Predator Avoidance in a Desert Ecosystem: Implications for the Conservation of Large Mammalian Herbivores." *Biological Conservation*, 25, no. 3 (March 1983): 193–208. https://doi.org/10.1016/0006-3207(83)90036-8.

Bitterman, M. E. "The Comparative Analysis of Learning." *Science*, 188, no. 4189 (May 1975): 699–709. https://doi.org/10.1126/science.188.4189.699.

Black, Henry Campbell. *Black's Law Dictionary: A Law Dictionary Containing Definitions of the Terms and Phrases of American and English Jurisprudence, Ancient and Modern*. 2nd ed. St Paul, MN: West Publishing Company, 1910. https://archive.org/details/blacks-law-dictionary-2nd-edition-1910/page/644/mode/2up.

Bohm, Robert T. "Artificial Nest Platforms for Raptors." *Raptor Research*, 11, no. 4 (Winter 1977): 97–99. https://sora.unm.edu/sites/default/files/journals/jrr/v011n04/p00097-p00099.pdf.

Bossert, William H., and Edward O. Wilson. "Analysis of Olfactory Communication Among Animals." *Journal of Theoretical Biology*, 5, no. 3 (November 1963): 443–469. https://doi.org/10.1016/0022-5193(63)90089-4.

Breck, Stewart W., Nathan Lance, and Victoria Seher. "Selective Foraging for Anthropogenic Resources by Black Bears: Minivans in Yosemite National Park." *Journal of Mammalogy*, 90, no. 5 (October 2009): 1041–1044. https://doi.org/10.1644/08-MAMM-A-056.1.

Bright, Michael. *100 Years of Wildlife*. London: BBC, 2007.

Buckner, C. H. "Preliminary Trials of a Camera Recording Device for the Study of Small Mammals." *Canadian Field Naturalist*, 78, no. 2 (1964): 77–79.

Bull, Evelyn L., Catherine G. Roberts, and Torolf R. Torgerson. "Trees and Logs Important to Wildlife in the Interior Columbia River Basin." *General Technical Report PNW-GTR-391*. Portland, OR: U.S. Department of Agriculture, Forest Service, Pacific Northwest Research Station, 1997. https://doi.org/10.2737/PNW-GTR-391.

Caro, Tim. *Antipredator Defenses in Birds and Mammals*. Chicago: University of Chicago Press, 2009.

Cassidy, Brandt G., and Robert A. Gonzales. "DNA Testing in Animal Forensics." *Journal of Wildlife Management*, 69, no. 4 (October 2005): 1454–1462. https://www.jstor.org/stable/3803505.

Champion, Frederick Walter. "Wildlife in the Indian Forests." Vol. 1 of *100 Years of Indian Forestry, 1861 to 1961*, edited by Dehra Dun. India: Forestry Research Institute, 1961.

Champion, Frederick Walter. *The Jungle in Sunlight and Shadow*. New York: Charles Scribner's Sons, 1934.

Champion, Frederick Walter. *Tripwire for a Tiger: Selected Works of F. W. Champion*. Chennai: Rainfed Books, 2012.

Champion, Frederick Walter. *With a Camera in Tiger-Land*. Garden City, NY: Doubleday, Doran & Company, 1928.

Chapman, Frank M. "Who Treads Our Trails? A Camera Trapper Describes His Experiences on an Island in the Canal Zone, a Natural-History Laboratory in the American Tropics." *National Geographic Magazine*, September 1927.

Chapman, Frank M. *My Tropical Air Castle*. New York: D. Appleton and Company, 1931.

Choegyal, Lisa. "Charles McDougal." *Nepali Times Buzz*, no. 808, May13–19, 2016. https://archive.nepalitimes.com/article/Nepali-Times-Buzz/Charles-McDougal-passes-away-at-86,3034.

Clapham, Melanie, Ed Miller, Mary Nguyen, and Chris T. Darimont. "Automated Facial Recognition for Wildlife that Lack Unique Markings: A Deep Learning Approach for Brown Bears." *Ecology and Evolution*, 10, no. 23 (December 2020): 12883–12892. https://doi.org/10.1002/ece3.6840.

Clark, Robert G., and Bruce K. Wobeser. "Making Sense of Scents: Effects of Odour on Survival of Simulated Duck Nests." *Journal of Avian Biology*, 28, no. 1 (March 1997): 31–37. https://doi.org/10.2307/3677091.

Coleman, Loren. "Debbie Martyr Rediscovers Extinct Muntjac." *CryptoZoo News*, October 10, 2008. www.cryptozoonews.com/martyr-muntjac/.

Craighead, John J., and Frank C. Craighead. *Hawks, Owls, and Wildlife*. New York: Dover Publications, 1969.

Curio, E. "The Adaptive Significance of Avian Mobbing: I. Teleonomic Hypotheses and Predictions." *Zeitschrift für Tierpsychologie*, 48, no. 2 (1978): 175–183.

Darwin, Charles. *The Expression of the Emotions in Man and Animals*. Chicago: University of Chicago Press, 1965.

Dawson, Mary, Laurent Marivaux, Chuan-Kui Li, K. Christopher Beard, and Grégoire Métais. "*Laonastes* and the 'Lazarus Effect' in Recent Mammals." *Science*, 311, no. 5766 (March 2006): 1456–1458. https://doi.org/10.1126/science.1124187.

Dinata, Yoan, Agung Nugroho, Iding Achmad Haidir, and Matthew Linkie. "Camera Trapping Rare and Threatened Avifauna in West-Central Sumatra." *Bird Conservation International*, 18, no. 1 (March 2008): 30–37. https://doi.org/10.1017/S0959270908000051.

Dobie, James Frank. *The Voice of the Coyote*. Lincoln: University of Nebraska Press, Bison Books, 1961.

Dodge, Wendell E., and Dana P. Snyder. "An Automatic Camera Device for Recording Wildlife Activity." *Journal of Wildlife Management*, 24, no. 3 (July 1960): 340–342. https://doi.org/10.2307/3797527.

Domjan, Michael. *The Principles of Learning and Behavior*. 4th ed. Pacific Grove, CA: Brooks/Cole, 1998.

Dugmore, Arthur Radclyffe. *The Romance of the Beaver: Being the History of the Beaver in the Western Hemisphere*. Philadelphia: J. B. Lippincott, 1914.

Ebersole, Rene. "Ethics: Too Close for Comfort." *Audubon Magazine*, May–June 2015. www.audubon.org/magazine/may-june-2015/too-close-comfort.

Edmunds, Malcolm. *Defence in Animals: A Survey of Anti-Predator Defences*. New York: Longman Group, 1974.

Estes, Richard D. "The Role of the Vomeronasal Organ in Mammalian Reproduction." *Mammalia*, 36, no. 3 (1972): 315-341. https://doi.org/10.1515/mamm.1972.36.3.315.

Fair, J. "Scientists Foxed by New Carnivore." *BBC Wildlife Magazine*, 24, no. 1 (2006): 30.

Flux, John E. C. "The Sumatran Rabbit *Nesolagus netscheri*." In *Rabbits, Hares and Pikas: Status, Survey and Conservation Action Plan*, edited by Joseph A. Chapman and John E. C. Flux. Gland, Switzerland: IUCN, 1990.

Forsman, Eric D., E. Charles Meslow, and Howard M. Wight. "Distribution and Biology of the Spotted Owl in Oregon." *Wildlife Monographs*, no. 87 (April 1984): 3–64. https://www.jstor.org/stable/3830695.

Forsman, Eric D., James K. Swingle, and Nicholas R. Hatch. "Behavior of Red Tree Voles (*Arborimus longicaudus*) Based on Continuous Video Monitoring of Nests." *Northwest Science*, 83, no. 3 (July 2009): 262–272. https://doi.org/10.3955/046.083.0309.

Frame, Paul F., H. Dean Cluff, and David S. Hik. "Response of Wolves to Experimental Disturbance at Homesites." *Journal of Wildlife Management*, 71, no. 2 (April 2007): 316–320. https://www.jstor.org/stable/4495188.

Freeland, W. J., and Daniel H. Janzen. "Strategies in Herbivory by Mammals: The Role of Plant Secondary Compounds." *The American Naturalist*, 108, no. 961 (May–June 1974): 269–289. https://doi.org/10.1086/282907.

French, Brett. "Photographer Captures Fight over Den between Badger, Fox." *Billings Gazette*. February 8, 2020. https://billingsgazette.com/lifestyles/recreation/photographer-captures-fight-over-den-between-badger-fox/article_ed072448-1c9c-5be4-9af2-4d246f50c8b0.html.

Frid, Alejandro. "Vigilance by Female Dall's Sheep: Interactions Between Predation Risk Factors." *Animal Behaviour*, 53, no. 4 (April 1997): 799–808. https://doi.org/10.1006/anbe.1996.0345.

Gabriel, Pia, and Richard T. Golightly. "Aversive Conditioning of Steller's Jays to Improve Marbled Murrelet Nest Survival." *Journal of Wildlife Management*, 79, no. 5 (July 2014): 894–903. https://doi.org/10.1002/jwmg.725.

Gackstetter, Chuck. *Trail Camera Trails of Lincoln County, Minnesota*. N.p: Chuck Gackstetter, 2012.

Gasset, José Ortega y. *Meditations on Hunting*. Translated by Howard B. Wescott. Belgrade MT: Wilderness Adventures Press, 1995.

Gipson, Philip S., Warren B. Ballard, and Ronald M. Nowak. "Famous North American Wolves and the Credibility of Early Wildlife Literature." *Wildlife Society Bulletin*, 26, no. 4 (Winter 1998): 808–816. https://www.jstor.org/stable/3783554.

Grand, Theodore I. *Carving Nature at Its Joints: Mammalian Anatomy, Behavior, Development, and Evolution*. Washington, DC: BRG Scientific, 2022.

Gregory, Tappan. "Lion in the Carmens." *The Chicago Naturalist* 1, no. 3–4 (1938): 73.

Gregory, Tappan. "The Black Wolf of the Tensas." *Program Activities of the Chicago Academy of Sciences*, 6, no. 3 (1935): 35–68.

Gregory, Tappan. *Eyes in the Night*. New York: Thomas Y Crowell Co., 1939.

Gregory, William King. *Biographical Memoir of Frank Michler Chapman, 1864–1945*. Washington, DC: National Academy of Sciences, 1947.

Grigonis, Hillary K. "'Spy in the Wild': BBC Documentary Hides Cameras Inside Robotic Animals." *Digital Trends*. January 12, 2017. https://www.digitaltrends.com/photography/robotic-animals-hidden-cameras/.

Gross, Charles G. *Donald R. Griffin 1915–2003: A Biographical Memoir*. Washington, DC: National Academies Press, 2005. http://www.nasonline.org/publications/biographical-memoirs/memoir-pdfs/griffin-donald.pdf.

Guynup, Sharon, and Nicolas Ruggia. "Owls Face Spotted Future." *National Geographic Channel*, July 22, 2004.

Guyton, Arthur C. *Textbook of Medical Physiology*. Philadelphia: W. B. Saunders Company, 1961.

Haagen-Smit, A.J. "Essential Oils: A Brief Survey of Their Chemistry and Production in the United States." *Economic Botany*, 3, no. 1 (January 1949): 71–83. https://doi.org/10.1007/BF02859506.

Habib, B., and S. Kumar. "Den Shifting by Wolves in Semi-Wild Landscapes in the Deccan Plateau, Maharashtra, India." *Journal of Zoology*, 272, no. 3 (July 2007): 259–265. https://doi.org/10.1111/j.1469-7998.2006.00265.x.

Hance, Jeremy. "The Saola: Rushing to Save the 'Most Spectacular Zoological Discovery' of the Twentieth Century." *Mongabay*, April 4, 2011. https://news.mongabay.com/2011/0404-hance_robichaud.html.

Hardy, Paul C., and Michael L. Morrison. "Factors Affecting the Detection of Elf Owls and Western Screech Owls." *Wildlife Society Bulletin*, 28, no. 2 (Summer 2000): 333–342. https://www.jstor.org/stable/3783689.

Harris, Charles E., and Frederick F. Knowlton. "Differential Responses of Coyotes to Novel Stimuli in Familiar and Unfamiliar Settings." *Canadian Journal of Zoology*, 79, no. 11 (November 2001): 2005–2013. https://doi.org/10.1139/z01-163.

Heinrich, Bernd. *The Mind of the Raven: Investigations and Adventures with Wolf-Birds*. New York: Harper Collins, 2000.

Hill, James O., Edward J. Pavlik, G. L. Smith III, Gordon M. Burghardt, and P. B. Coulson. "Species-Characteristic Responses to Catnip by Undomesticated Felids." *Journal of Chemical Ecology*, 2, no. 2 (1976): 239–253. https://doi.org/10.1007/BF00987747.

Holden, Constance. "New Species in Borneo?" *Science*, 310, no. 5755 (December 2005): 1764. https://doi.org/10.1126/science.310.5755.1764d.

Huchon, Dorothée, Pascale Chevret, Ursula Jordan, C. William Kilpatrick, Vincent Ranwez, Paulina D. Jenkins, Jürgen Brosius, and Jürgen Schmitz. "Multiple Molecular Evidences for a Living Mammalian Fossil." *Proceedings of the National Academy of Science*, 104, no. 18 (May 2007): 7495–7499. https://doi.org/10.1073/pnas.0701289104.

Jenkins, Paulina D., C. William Kilpatrick, Mark F. Robinson, and Robert J. Timmins. "Morphological and Molecular Investigations of a New Family, Genus and Species of Rodent (Mammalia: Rodentia: Hystricognatha) from Lao PDR." *Systematics and Biodiversity*, 2, no. 4 (2005): 419–454.

Jobanek, George A. "Searching for the Tree Vole: An Episode in the 1914 Biological Survey of Oregon." *Oregon Historical Quarterly*, 89, no. 4 (Winter 1988): 369–400.

Karanth, K. Ullas, and James D. Nichols, eds. *Methods for Monitoring Tiger and Prey Populations*. Singapore: Springer Nature, 2017. https://doi.org/10.1007/978-981-10-5436-5.

Karanth, K. Ullas, and James D. Nichols. "Estimation of Tiger Densities in India Using Photographic Captures and Recaptures." *Ecology*, 79, no. 8 (December 1, 1998): 2852–2862. https://doi.org/10.1890/0012-9658(1998)079[2852:EOTDII]2.0.CO;2.

Karanth, K. Ullas. "Estimating Tiger *Panthera tigris* Populations from Camera-Trap Data Using Capture–Recapture Models." *Biological Conservation*, 71, no. 3 (1995): 333–338. https://doi.org/10.1016/0006-3207(94)00057-W.

Karanth, K. Ullas. "The Trouble with Tiger Numbers." *Scientific American*, (July 1, 2016): 55–59. www.scientificamerican.com/article/the-trouble-with-tiger-numbers/.

Kays, Roland, Monica Lasky, Maximilian L. Allen, Robert Dowler, Melissa Hawkins, Andrew Hope, Brooks Kohli, et al. "Which Mammals Can Be Identified from Camera Traps and Crowdsourced Photographs?" *Journal of Mammalogy*, 103, no. 4 (August 2022): 767–775. https://doi.org/10.1093/jmammal/gyac021.

Kloppers, Elsabé Louise, Colleen Cassady St. Clair, and Thomas Eric Hurd. "Predator-Resembling Aversive Conditioning for Managing Habituated Wildlife." *Ecology and Society*, 10, no. 1 (2005): 442–450. www.ecologyandsociety.org/vol10/iss1/art31/.

Knörnschild, Mirjam, and Marco Tschapka. "Predator Mobbing Behaviour in the Greater Spear-Nosed Bat, *Phyllostomus hastatus*." *Chiroptera Neotropical*, 18, no. 2 (December 2012): 1132–1135.

Kucera, Thomas E., and Reginald H. Barrett. "A History of Camera Trapping." In *Camera Traps in Animal Ecology: Methods and Analyses*, edited by Allan F. O'Connell, James D. Nichols, and K. Ullas Karanth. Tokyo: Springer, 2011. https://doi.org/10.1007/978-4-431-99495-4.

Kucera, Thomas E., and Reginald H. Barrett. "A History of Camera Trapping." In *Camera Traps in Animal Ecology: Methods and Analyses,* edited by A.F. O'Connell, J.D. Nichols, and U.K. Karanth. Tokyo: Springer, 2011.

Larrucea, Eveline S., and Peter F. Brussard. "Habitat Selection and Current Distribution of the Pygmy Rabbit in Nevada and California, USA." *Journal of Mammalogy*, 89, no. 3 (June 2008): 691–699. https://doi.org/10.1644/07-MAMM-A-199R.1.

Lasley, Bill L., and Jay F. Kirkpatrick. "Monitoring Ovarian Function in Captive and Free-Ranging Wildlife by Means of Urinary and Fecal Steroids." *Journal of Zoo and Wildlife Medicine*, 22, no. 1 (March 1991): 23–31. https://www.jstor.org/stable/20095115.

Lehner, Phillip N. "Coyote Vocalizations: A Lexicon and Comparisons with Other Canids." *Animal Behaviour*, 26, part 3 (August 1978): 712–722. https://doi.org/10.1016/0003-3472(78)90138-0.

Leopold, Aldo. *A Sand County Almanac*. Oxford: Oxford University Press, 1949.

Linderman, Frank B. *Plenty-Coups: Chief of the Crows*. Lincoln: University of Nebraska Press, 2002.

Linnell, John D., Jon E. Swenson, Reidar Andersen, and Brian Barnes. "How Vulnerable Are Denning Bears to Disturbance?" *Wildlife Society Bulletin*, 28, no. 2 (Summer 2000): 400–413. https://www.jstor.org/stable/3783698.

Majid, Asifa. "Human Olfaction at the Intersection of Language, Culture, and Biology." *Trends in Cognitive Sciences*, 25, no. 2 (February 2021): 111–123. https://doi.org/10.1016/j.tics.2020.11.005.

Mariani, Evi. "Unexplored Forest Reveals Mystery." *Jakarta Post*, December 9, 2005.

Maser, Chris. *Mammals of the Pacific Northwest: From the Coast to the High Cascades*. Corvallis: Oregon State University Press, 1998.

Mazur, Rachel, and Victoria Seher. "Socially Learned Foraging Behaviour in Wild Black Bears, *Ursus americanus*." *Animal Behaviour*, 75, no. 4 (April 2008): 1503–1508. http://dx.doi.org/10.1016/j.anbehav.2007.10.027.

Mazur, Rachel. "Does Aversive Conditioning Reduce Human–Black Bear Conflict?" *Journal of Wildlife Management*, 74, no. 1 (January 2010): 48–54. https://www.jstor.org/stable/20616828.

McCarthy, Jennifer, Todd K. Fuller, Kyle P. McCarthy, Hariyo Wibisono, and Mark C. Livolsi. "Using Camera Trap Photos and Direct Sightings to Identify Possible Refugia for the Vulnerable Sumatran Striped Rabbit *Nesolagus netscheri*." *Oryx*, 46, no. 3 (May 3, 2012): 438–441. https://doi.org/10.1017/S0030605312000051.

McCullough, Dale R. "Behavior, Bears, and Humans." *Wildlife Society Bulletin*, 10, no. 1 (Spring 1982): 27–33. https://www.jstor.org/stable/3781798.

McDaniel, Gregory W., Kevin S. McKelvey, John R. Squires, and Leonard F. Ruggiero. "Efficacy of Lures and Hair Snares to Detect Lynx." *Wildlife Society Bulletin*, 28, no. 1 (2000): 119–123.

McNally, Joe. *Hot Shoe Diaries: Big Light from Small Flashes*. Berkeley: New Riders, 2009.

Meijaard, Erik, Andrew C. Kitchener, and Chris Smeenk. "'New Bornean Carnivore' Is Most Likely a Little Known Flying Squirrel." *Mammal Review*, 36, no. 4 (December 2006): 318–324. https://doi.org/10.1111/j.1365-2907.2006.00089.x.

Meiselman, N., and A.T. Doyle. "Habitat and Microhabitat Use by the Red Tree Vole (*Phenacomys longicaudus*)." *American Midland Naturalist*, 135, no. 1 (January 1996): 33–42. https://doi.org/10.2307/2426869.

Mettler, Alex Edward, and John Anthony Shivik. "Dominance and Neophobia in Coyote (*Canis latrans*) Breeding Pairs." *Applied Animal Behaviour Science*, 102, no. 1–2 (January 2007): 85–94. https://doi.org/10.1016/j.applanim.2006.03.012.

Mirsky, Steve. "Kha-nyou Smell a Rat? Tracking Down Real Rock Rodents and Imaginary Amazon Apes." *Scientific American*, August 1, 2005. ww.scientificamerican.com/article/kha-nyou-smell-a-rat/.

Montana Fish, Wildlife, and Parks. *Deer, Elk, and Antelope Hunting Regulations*. Montana FWP, 2022. https://fwp.mt.gov/binaries/content/assets/fwp/hunt/regulations/2022/2022-dea-regulations-final-for-web.pdf.

Morrell, Thomas E., Richard H. Yahner, and William L. Harkness. "Factors Affecting Detection of Great Horned Owls by Using Broadcast Vocalizations." *Wildlife Society Bulletin*, 19, no. 4 (1991): 481–88. http://www.jstor.org/stable/3782162.

Morrissey, Rory. *Oliver G. Pike 1877–1963: Leighton Buzzard Naturalist, Author, Photographer*. Bedfordshire: Leighton Buzzard and District Archaeological and Historical Society, 2013.

Muir, John A. "Windstorm in the Forests." *The Mountains of California*. N.p, 1894. https://vault.sierraclub.org/john_muir_exhibit/writings/the_mountains_of_california/chapter_10.aspx.

Müller Schwarze, Dietland, and Lixing Sun. *The Beaver: Natural History of a Wetlands Engineer*. Ithaca, NY: Cornell University Press, 2003.

Murie, Adolph. *Ecology of the Coyote in the Yellowstone: Fauna of the National Parks of the United States*. Washington, DC: National Park Service, 1940.

Nams, V. "Density-Dependent Predation by Skunks Using Olfactory Search Images." *Oecologia*, 110 (April 1997): 440–448. https://doi.org/10.1007/s004420050179.

Newbold, Haylie G., and Carolyn M. King. "Can a Predator See 'Invisible' Light? Infrared Vision in Ferrets (*Mustela furo*)." *Wildlife Research*, 36, no. 4 (June 2009): 309–318. https://doi.org/10.1071/WR08083.

Nguyen, An, Andrew Tilker, Duy Le, Huong Van Le, Son Van Le, Truong Hong Luu, Bang Van Tran, and Andreas Wilting. "New Records and Southern Range Extension of the Annamite Striped Rabbit *Nesolagus timminsi* in Vietnam." *Mammalia*, 85, no. 5 (2021): 417–421. https://doi.org/10.1515/mammalia-2020-0189.

Nguyen, An, Van Bang Tran, Duc Minh Hoang, Thi Anh Minh Nguyen, Dinh Thang Nguyen, Van Tiep Tran, Barney Long, Erik Meijaard, Jeff Holland, Andreas Wilting, and Andrew Tilker. "Camera-Trap Evidence That the Silver-Backed Chevrotain *Tragulus versicolor* Remains in the Wild in Vietnam." *Nature Ecology & Evolution*, 3, (November 2019): 1650–1654. https://doi.org/10.1038/s41559-019-1027-7.

Nonaka, Yuki. *Response of Breeding Wolves to Human Disturbance on Den Sites: An Experiment*. Sweden: Biology Education Centre, Uppsala University, 2011. https://www.slu.se/globalassets/ew/org/inst/ekol/forskning/projekt/skandulv/publikationer/studentarbeten/nonaka-2011response-of-breeding-wolves-to-human-disturbance-on-den-sites--an-experiment.pdf.

Olszewski, Jerzy. "Role of Uprooted Trees in the Movements of Rodents in Forests." *Oikos*, 19, no. 1 (1968): 99–104.

Osterberg, Donald M. "Activity of Small Mammals as Recorded by a Photographic Device." *Journal of Mammalogy*, 43, no. 2 (May 29, 1962): 219–229. https://doi.org/10.2307/1377093.

Owen, James. "Photos: Cameras 'Trap' Hairy-Nosed Otter, More Rarities." *National Geographic*, July 26, 2010. https://www.nationalgeographic.com/animals/article/100726-hairy-nosed-otter-borneo-camera-trap-dera-makot-science-pictures.

Palmer, Chris. *Shooting in the Wild: An Insider's Account of Making Movies in the Animal Kingdom*. San Francisco: Sierra Club Books, 2010.

Palminteri, Sue. "Camera Trapping in the Trees." *Mongabay*. July 4, 2017. https://news.mongabay.com/wildtech/2017/07/camera-trapping-in-the-trees/.

Pearson, Oliver P. "A Traffic Survey of *Microtus-Reithrodontomys* Runways." *Journal of Mammalogy*, 40, no. 2 (1959): 169–180. https://doi.org/10.2307/1376431.

Pearson, Oliver P. "Habits of Harvest Mice Revealed by Automatic Photographic Recorders." *Journal of Mammalogy*, 41, no. 1 (1960): 58–74.

Pearson, Oliver P. "Habits of *Microtus californicus* Revealed by Automatic Photograph Records." *Ecological Monographs*, 30 (1960): 231–249.

Peeters, Hans. *Field Guide to Owls of California and the West*. Berkeley: University of California Press, 2007.

Randler, Christoph, Tobias Katzmaier, Jochen Kalb, Nadine Kalb, and Thomas Gottschalk. "Baiting/Luring Improves Detection Probability and Species Identification—A Case Study of Mustelids with Camera Traps." *Animals*, 10, no. 11 (November 2020): 2178–2186. https://doi.org/10.3390/ani10112178.

Raphael, Martin G., and Marshall White. "Use of Snags by Cavity-Nesting Birds in the Sierra Nevada." *Wildlife Monographs*, no. 86 (1984): 1–66. https://www.jstor.org/stable/3830575.

Ray, D. G. H. *A Portable Automatic Camera System for Recording Animal Activity*. Winnipeg, Manitoba: Northern Forestry Centre, 1967.

Reed, Sarah E., and Adina M. Merenlender. "Quiet, Nonconsumptive Recreation Reduces Protected Area Effectiveness." *Conservation Letters*, 1, no. 3 (August 2008): 146–154. https://doi.org/10.1111/j.1755-263X.2008.00019.x.

Reynolds, H. C. "Studies of Reproduction in the Opossum (*Didelphis virginiana virginiana*)." *University of California Publications in Zoology*, 52, no. 3 (1952): 223–284.

Roberts, Claude. "Sniffer Dogs: The World of K9 Scent Detection." *Dog Learner*. July 24, 2020.

Robichaud, William G. "Physical and Behavioral Description of a Captive Saola, *Pseudoryx nghetinhensis*." *Journal of Mammalogy*, 79, no. 2 (May 20, 1998): 394–405. https://doi.org/10.2307/1382970.

Romer, Alfred Sherwood. *The Vertebrate Body*. Philadelphia: W. B. Saunders Company, 1962.

Root, Alan. *Ivory, Apes and Peacocks: Animals, Adventures and Discovery in the Wild Places of Africa*. London: Vintage Books, 2013.

Rosell, Frank, and JørnIngar Sanda. "Potential Risks of Olfactory Signaling: The Effect of Predators on Scent Marking by Beavers." *Behavioral Ecology*, 17, no. 6, (November 2006): 897–904. https://doi.org/10.1093/beheco/arl022.

Rovero, Francesco, and Fridolin Zimmermann, eds. *Camera Trapping for Wildlife Research*. Exeter, UK: Pelagic Publishing, 2016.

Rovero, Francesco, and Galen B. Rathbun. "A Potentially New Giant Sengi (Elephant-Shrew) from the Udzungwa Mountains, Tanzania." *Journal of East African Natural History*, 95, no. 2 (July 1, 2006): 111–115. https://doi.org/10.2982/0012-8317(2006)95[111:APNGSE]2.0.CO;2.

Rovero, Francesco, Galen Rathbun, Andrew Perkin, Trevor Jones, David Ribble, Charles Leonard, Ruben Mwakisoma, and Nike Doggart. "A New Species of Giant Sengi or Elephant-Shrew (Genus *Rhynchocyon*) Highlights the Exceptional Biodiversity of the Udzungwa Mountains of Tanzania." *Journal of Zoology*, 274, no. 2 (January 25, 2008): 126–133. https://doi.org/10.1111/j.1469-7998.2007.00363.x.

Runte, Alfred. *Yosemite: The Embattled Wilderness*. Lincoln: University of Nebraska Press, 1990.

Russell, Carl P. *Firearms, Traps, and Tools of the Mountain Men*. Albuquerque: University of New Mexico Press, 1967.

Russell, Osborne. *Journal of a Trapper: Nine Years in the Rocky Mountains 1834–1843*. Lincoln: University of Nebraska Press, 1965.

Rydell, Jens, Danilo Russo, Price Sewell, Ernest C. J, Seamark, Charles M. Francis, Sherri L. Fenton, and M. Brock Fenton. "Bat selfies: Photographic Surveys of Flying Bats," *Mammalian Biology*, 102, no. 3 (April 7, 2022): 793–809. https://link.springer.com/article/10.1007/s42991-022-00233-7.

Sandoz, Maro. *The Beaver Men: Spearheads of Empire*. Lincoln: University of Nebraska Press, 1978.

Schillings, Carl Georg. *With Flashlight and Rifle: Photographing by Flash-Light at Night the Wild Animal World of Equatorial Africa*. New York: Harper & Brothers Publishers, 1905.

Schlexer, Fredrick V. "Attracting Animals to Detection Devices." In *Noninvasive Survey Methods for Carnivores*, edited by Robert A. Long, Paula MacKay, Justina Ray, and William Zielinski. Washington, DC: Island Press, 2008.

Seidman, David, and Paul Cleveland. *The Essential Wilderness Navigator: How to Find Your Way in the Great Outdoors*. 2nd ed. New York: Ragged Mountain Press, McGraw Hill, 2001.

Sequin, Eveline S., Michael M. Jaeger, Peter F. Brussard, and Reginald H. Barrett. "Wariness of Coyotes to Camera Traps Relative to Social Status and Territory Boundaries." *Canadian Journal of Zoology*, 81, no. 12 (December 2003): 2015–2025. https://doi.org/10.1139/z03-204.

Shiras, George. *Hunting Wildlife with Camera and Flashlight*. Vol 1. Washington, DC: National Geographic Society, 1936.

Smith, James L. David, Charles McDougal, Sean C. Ahearn, Anup Joshi, and Kathy Conforti. "Metapopulation Structure of Tigers in Nepal." In *Riding the Tiger: Tiger Conservation in Human-Dominated Landscapes,* edited by John Seidensticker, Sarah Christie, and Peter Jackson. Cambridge: Cambridge University Press, 1999.

Smythe, Nicholas. *The Natural History of the Central American Agouti* (Dasyprocta punctata). Washington, DC: Smithsonian Institution Press, 1978. https://doi.org/10.5479/si.00810282.257.

Stager, Kenneth. "The Role of Olfaction in Food Location by the Turkey Vulture (*Cathartes aura*)." *Contributions in Science,* no. 81, June 30, 1964.

Stalmaster, Mark V., and James L. Kaiser. "Effects of Recreational Activity on Wintering Bald Eagles." *Wildlife Monographs*, no. 137 (April 1998): 1–46. https://www.jstor.org/stable/3830779.

Steidl, Robert J., and Robert G. Anthony. "Responses of Bald Eagles to Human Activity During the Summer in Interior Alaska." *Ecological Applications*, 6, no. 2 (May 1996): 482–491. https://doi.org/10.2307/2269385.

Stephenson, Frank. "Retired FSU Professor Captures a 'Living Fossil' Laotian Rock Rat on Video." *Florida State University News*, June 13, 2006. https://www.fsu.edu/news/2006/06/13/rock.rat/.

Stockman, Steve. *How to Shoot Video That Doesn't Suck: Advice to Make Any Amateur Look Like a Pro*. New York: Workman Publishing, 2011.

Struhsaker, Thomas T. "Behavior of Elk (*Cervus canadensis*) During the Rut." *Zeitschrift für Tierpsychologie*, 24, no. 1 (January–December 1967): 80–114. https://doi.org/10.1111/j.1439-0310.1967.tb01229.x.

Stümpke, Harald. *The Snouters: Form and Life of the Rhinogrades*. Translated by Leigh Chadwick. Chicago: University of Chicago Press, 1981.

Surridge, Allison K., Robert J. Timmins, Godfrey M. Hewitt, and Diana J. Bell. "Striped Rabbits in Southeast Asia." *Nature*, 400 (August 19, 1999): 726. https://doi.org/10.1038/23393.

Svendsen, Gerald E., and Kenneth B. Armitage. "Mirror-Image Stimulation Applied to Field Behavioral Studies." *Ecology*, 54, no. 3 (May 1973): 623–627. https://doi.org/10.2307/1935349.

The Nature Group. *Nature Photographers' Code of Practice*. The Royal Photographic Society, 2007. https://rps.org/media/1xcnsuga/the-nature-photographers-code-of-practice.pdf.

Thiel, Richard P., Samuel Merrill, and L. David Mech. "Tolerance by Denning Wolves, *Canis lupus*, to Human Disturbance." *Canadian Field Naturalist*, 112, no. 2 (April–June 1998): 340–342.

Thomsen, Phillip F., and Eske Willerslev. "Environmental DNA: An Emerging Tool in Conservation for Monitoring Past and Present Biodiversity." *Biological Conservation*, 183 (March 2015): 4–18. https://doi.org/10.1016/j.biocon.2014.11.019.

Thomson, Robert, Gail C. Potgieter, and Laila Bahaa-el-din. "Closing the Gap Between Camera Trap Software Development and the User Community." *African Journal of Ecology*, 56, no. 4 (December 2018): 721–739. https://doi.org/10.1111/aje.12550.

Thorn, Michelle, Dawn M. Scott, Matthew Green, Phillip W. Bateman, and Elissa Z. Cameron. "Estimating Brown Hyaena Occupancy Using Baited Camera Traps." *South African Journal of Wildlife Research*, 39, no. 1 (April 2009): 1–10. http://dx.doi.org/10.3957/056.039.0101.

Tilker, Andrew. "Fading Stripes in Southeast Asia: Saving the Elusive Annamite Striped Rabbit." *Zoology*, December 18, 2018. https://beta.capeia.com/zoology/2018/12/13/fading-stripes-in-southeast-asia-saving-the-elusive-annamite-striped-rabbit.

Tinbergen, Niko. *Curious Naturalists*. New York: Doubleday Anchor/American Museum of Natural History, 1968.

Tobler, Mathias, S. E. Carillo-Pergastigui, R. Leite Pitman, R. Mares, and G. Powell. "An Evaluation of Camera Traps for Inventorying Large- and Medium-Sized Terrestrial Rainforest Mammals." *Animal Conservation*, 11, no. 3 (June 2008): 169–178. https://doi.org/10.1111/j.1469-1795.2008.00169.x.

Turkowski, Frank J., Martin L. Popelka, and Roger W. Bullard. "Efficacy of Odor Lures and Baits for Coyotes." *Wildlife Society Bulletin*, 11, no. 2 (1983): 136–145.

Turner, R. "How Does Catnip Work Its Magic on Cats?" *Scientific American*, May 29, 2007. https://www.scientificamerican.com/article/experts-how-does-catnip-work-on-cats/.

Tye, Simon P., Keith Geluso, Mary J. Harner, Adam M. Siepielski, Michael L. Forsberg, Emma M. Brinley Buckley, Jeffrey S. Dale. "One House Is a Home for Many: Temporal Partitioning of Vertebrates on an American Beaver Lodge." *American Midland Naturalist*, 185, no. 2 (April 2021): 229–240. https://doi.org/10.1674/0003-0031-185.2.229.

Verderber, Gustav W. "Camera Trapping: How to Get the Shot." *Northern Woodlands*, Winter 2011: 34. https://northernwoodlands.org/articles/article/camera-trapping-how-to-get-the-shot/.

Verts, B. J. *The Biology of the Striped Skunk*. Urbana: University of Illinois Press, 1967.

Viitala, Jussi, Erkki Korpimäki, Minna Koivula, and Pälvi Palokangas. "Attraction of Kestrels to Vole Scent Marks Visible in Ultraviolet Light." *Nature*, 373 (February 1995): 423–425. https://doi.org/10.1038/373425a0.

Vu Van Dung, and Du Tuoc. "The Discovery of the Saola (*Pseudoryx nghetinhensis*) in Vietnam." In *Rediscovering the Saola: A Status Review and Conservation Planning Workshop Proceedings*, edited by J. Hardcastle, et al., Hanoi: WWF Indochina Programme, 2004. https://wwfeu.awsassets.panda.org/downloads/saolaproceedingenglish.pdf.

Vu Van Dung, Pham Mong Giao, Nguyen Ngoc Chinh, Do Tuoc, Peter Arctander, and John Mackinnon. "A New Species of Living Bovid from Vietnam." *Nature*, 363, (June 3, 1993): 443–445. https://doi.org/10.1038/363443a0.

Wegge, Per, Chiranjibi Pd. Pokheral, and Shant Raj Jnawali. "Effects of Trapping Effort and Trap Shyness on Estimates of Tiger Abundance from Camera Trap Studies." *Animal Conservation*, 7, no. 3 (August 2004): 251–256. https://doi.org/10.1017/S1367943004001441.

Williams, Ted. "Phony Wildlife Gives a Warped Sense of Nature." *Audubon Magazine*, March–April 2010. www.audubonmagazine.org/articles/nature/phony-wildlife-photography-gives-warped-view-nature.

Windberg, Lamar A. "Coyote Responses to Visual and Olfactory Stimuli Related to Familiarity with an Area." *Canadian Journal of Zoology*, 74, no. 12 (December 1996): 2248–2253. https://doi.org/10.1139/z96-255.

Winters, Loren A. *A Tutorial in High-Speed Photography with Electronic Flash*. Self-pub., 1992–1996. https://www.hiviz.com/activities/guidebook/hsptutor.pdf.

Young, Stanley P. *Sketches of American Wildlife*. Baltimore: Monumental Press, 1946.

Young, Stanley P. *The Last of the Loners*. New York: MacMillan, 1970.

Young, Stanley P., and Harold W. Dobyns. *Den Hunting as a Means of Coyote Control: US Department of Agriculture Leaflet No. 132*. Washington, DC: US Government Printing Office, 1937.

Websites, Blogs, and Videos

Agarwal, Amit. "Tools for Managing EXIF Data of Your Images." *Digital Inspiration* (blog). May 20, 2012. https://www.labnol.org/software/exif-data-editors/14210/.

American Society of Mammalogists. "Mammal Images Library." American Society of Mammalogists (website). Accessed December 6, 2022. https://www.mammalsociety.org/committees/mammal-images-library.

Amrhein, Jim. "Animal Rights vs. Human Rights: 'Righting' a Wrong." Daily Reckoning (website). Accessed December 5, 2022. https://dailyreckoning.com/animal-rights-vs-human-rights/.

Anand, Sweta. "24 Best Free EXIF Viewer Software for Windows." List of Freeware (website). Accessed December 6, 2022. https://listoffreeware.com/free-exif-viewer-software-.indows/.

Anon. "Beim Tresorknacken im Wald gefilmt." [Filmed Cracking the Safe in the Woods]. *Burgenland News*. March 20, 2008. https://bglv1.orf.at/stories/264976.

ASCPF. "International Federation of Wildlife and Nature Photography." ASCPF (website). Accessed December 5, 2022. https://www.ascpf.fr/ifwp/.

Ballotpedia. "Endangered Species by State." Ballotpedia (website). Accessed December 6, 2022. https://ballotpedia.org/Endangered_species_by_state.

Banisar, David, and Simon Davies. "Privacy and Human Rights, An International Survey of Privacy Laws and Practice." Global Internet Liberty Campaign (website). Accessed December 5, 2022. https://gilc.org/privacy/survey/intro.html.

Benson, Etienne. "Paparazzi in the Woods: Hidden Surveillance Cameras Are Making the Wilderness Less Wild." *Slate* (website). August 14, 2008. https://slate.com/technology/2008/08/camera-traps-hidden-in-the-woods-are-making-animals-angry.html.

Bigboyz [pseud.]. "Trail Camera Anti-Theft." HuntingNet.com (forum). June 23, 2005. https://www.huntingnet.com/forum/bowhunting/103633-trail-camera-anti-theft.html.

Boot, Michael. "Bears vs. Bees—Nature's Secret Power." YouTube. August 17, 2012. https://www.youtube.com/watch?v=YJ42pMtmrlE.

Booth, Robert. "Wildlife Photographer of the Year Stripped of His Award." *The Guardian* (website). January 20, 2010. https://www.theguardian.com/environment/2010/jan/20/wolf-wildlife-photographer-award-stripped.

Bright Hub. "What is a Slave Flash & Tips on How to Use a Slave Flash." Bright Hub (website). May 17, 2010. https://www.brighthub.com/multimedia/photography/articles/71464/.

Brown, Matt. "Mountain Lion Steals Game Caller." YouTube. November 12, 2013. https://www.youtube.com/watch?v=Afw5K7MYkbo.

Burrard-Lucas, William. "Turn your DSLR Camera into a Wildlife Camera Trap." Digital Photography School (website). Accessed December 6, 2022. https://digital-photography-school.com/turn-dslr-camera-wildlife-camera-trap/.

Cadle, Jake. "Upload GPS Coordinates to Google Earth Pro." YouTube. January 25, 2016. https://youtu.be/1HgDN-cyebJo.

Californian Nevada River Forecast Center. "Observed Precipitation." NOAA (website). Accessed December 2, 2022. https://www.cnrfc.noaa.gov/rainfall_data.php.

Cambush. "Do It Yourself Trail Camera Mount." *Cambush* (blog). October 12, 2013. https://cambushcamo.com/blogs/news/9586237-do-it-yourself-trail-camera-mount.

Carwardine, Mark. "Ethics in Wildlife Photography: The Good, the Bad, and the Ugly." *Discover Wildlife* (website). November 7, 2011. https://web.archive.org/web/20171228155938/https://www.discoverwildlife.com/wildlife-nature-photography/ethics-wildlife-photography?page=3.

Casey, Rachel. "'Learning Theory' Terminology: Classical and Operant Conditioning." *Raining Cats and Dogs* (blog). October 12, 2013. https://behaviourvet.wordpress.com/2013/10/12/learning-theory-terminology-classical-and-operant-conditioning/.

Cobb, Robert. "Homemade E-Caller." Predator Masters (forum). Accessed December 6, 2022. https://www.predatormastersforums.com/homemadeecaller.shtml.

Cooper, Spence. "Jamie Oliver Talks About Beaver Castoreum in Vanilla Ice Cream." *Friends Eat* (blog). April 12, 2011. https://blog.friendseat.com/jamie-oliver-beaver-castoreum-extract-letterman/.

Cornell Lab of Ornithology. "Platform for Success." NestWatch (website). Accessed December 1, 2022. https://nestwatch.org/connect/news/platform-for-success/.

Cornell Ornithology Lab. "All Cams." All About Birds. Accessed December 1, 2022. https://cams.allaboutbirds.org/all-cams/.

Cornpatcher [psued.]. "What Are Your Most Creative Ways to Hide Your Trail Cameras from Trespassers?" Archery Talk (forum). June 28, 2012. https://www.archerytalk.com/vb/showthread.php?t=1785899.

Creative Commons. "About." Creative Commons (website). Accessed December 6, 2022. https://creativecommons.org/about/.

Cremer, Jeff. "Termites Build a Nest in My Camera." *Camtrapper* (blog). September 4, 2015. https://camtrapper.com/viewtopic.php?f=7&t=9877#p80178.

Dale, Jeff. "All in a Days Work." *Camtrapper* (blog). April 22, 2016. https://camtrapper.com/viewtopic.php?f=13&t=8637&p=85786&hilit=TRLcam#p85786.

Dale, Jeff. "The DSLR and Mirrorless Camera List." *Camtrapper* (blog). April 4, 2014. https://camtrapper.com/viewtopic.php?f=113&t=8365.

Dale, Jeff. "TRLcam's Economy DSLR Camera Trap Build: Part One." *Camtrapper* (blog). March 12, 2012. https://camtrapper.com/viewtopic.php?f=113&t=5298#p38913.

Davexx1 [pseud.]. "Camo/Hide Your Trail Camera . . ." Georgia Outdoor News (forum). July 31, 2007. https://forum.gon.com/showthread.php?t=127217.

Digital Naturalism Laboratories. "Mysterious Castles of Clay." YouTube. November 24, 2013. https://www.youtube.com/watch?v=PLVJY31EUEE.

Easterson, Sam. "Burrow-Cams." Video Data Bank (website). Accessed December 1, 2022. https://www.vdb.org/titles/burrow-cams.

Estate of Oliver G Pike. "Oliver G. Pike." Oliver G Pike (website). Accessed December 6, 2022, http://www.olivergpike.info/.

Farmers Insurance. "Close Claws." Aban Commercials (website). Accessed February 7, 2023. https://abancommercials.com/farmers-insurance/hall-claims-close-claws-commercial/48056/.

Farrell, Kevin. "You Might Have Consumed This Beaver Sac Secretion without Knowing." *USA Today 10Best.com* (website). April 25, 2020. https://www.10best.com/interests/food-culture/beaver-sac-excretions-have-flavored-more-food-than-you-realize/.

Forestry Service. "Know Before You Go: Wildlife Photography." USDA (website). Accessed December 6, 2022. https://www.fs.usda.gov/visit/know-before-you-go/wildlife-photography.

Game Cam Wildlife. "How to Add an External Battery Mod to Your Game Trail Camera." YouTube. February 6, 2021. https://www.youtube.com/watch?v=9No_qXZ7Ftc.

Gannon, Meg. "No Pictures, Please! Andean Bears Wreck Research Camera." Yahoo News (website). October 23, 2013. https://news.yahoo.com/no-pictures-please-andean-bears-wreck-research-camera-135914153.html.

Garrett, Charles. "Vivitar 2800 Hack." YouTube. June 27, 2012. https://www.youtube.com/watch?v=fFVVGvl7UNg.

Graphic Converter. "Graphic Converter for Mac." Graphic Converter (website). Accessed December 6, 2022. https://www.graphic-converter.net/.

Hall, Travis. "What's the Future of Trail Cameras in the West?" The Meateater (website). January 20, 2022. https://www.themeateater.com/conservation/policy-and-legislation/whats-the-future-of-trail-cameras-in-the-west.

Hart, Michael. "History and Philosophy of Project Gutenberg." Project Gutenberg (website). Accessed December 6, 2022. https://www.gutenberg.org/about/background/history_and_philosophy.html.

Hart, Terese. "Eight Months in the Forest Canopy." *Searching for Bonobo in Congo* (blog). September 19, 2017. https://www.bonoboincongo.com/2017/09/19/eight-months-in-the-forest-canopy/.

Hickman, Ken. "I'm on a Perch." *Nature of a Man* (blog). August 5, 2012. https://natureofaman.blogspot.com/2012/08/im-on-perch.html.

Hickman, Ken. "Pygmies of the Great Basin." *Nature of a Man* (blog). July 23, 2014. https://natureofaman.blogspot.com/2014/07/pygmies-of-great-basin.html.

Hickman, Ken. "Rocky Relationships." *Nature of a Man* (blog). October 22, 2015. https://natureofaman.blogspot.com/2015/10/rocky-relationships.html.

Hickman, Ken. "Stephen Colbert's Nightmare." *Nature of a Man* (blog). October 29, 2011. https://natureofaman.blogspot.com/2011/10/stephen-colberts-nightmare.html.

Hickman, Ken. "The Coast Packrats." *Nature of a Man* (blog). January 25, 2012. https://natureofaman.blogspot.com/2012/01/coast-packrats.html.

Hindy, Joe. "10 Best Excel Apps and Spreadsheet Apps for Android." Android Authority (website). April 29, 2022. https://www.androidauthority.com/best-excel-apps-spreadsheet-apps-android-1122472/.

Hinton, Kyle. "Stolen Trail Camera Leads to Arrest of Thief." KFVS (website). May 12, 2018. https://www.kfvs12.com/story/38176808/stolen-trail-camera-leads-to-arrest-of-thief/.

Hobby, David. The Strobist (website). Accessed December 6, 2022. https://strobist.blogspot.com.

Hunting Farmer. "DIY Solar Trail Camera Power: External 12V Battery and Charger." YouTube. February 20, 2019. https://www.youtube.com/watch?v=PNOOY3cPPoo.

iNaturalist. "Help." iNaturalist (website). Accessed December 6, 2022. https://www.inaturalist.org/pages/help.

International Press Telecommunications Council. "What Is Photo Metadata?" IPTC (website). Accessed December 6, 2022. https://iptc.org/standards/photo-metadata/photo-metadata/.

Jack the Hat Photographic. "Off-Camera Flash Techniques Using Cheap Optical Slave Trigger Units." YouTube. September 27, 2012. https://www.youtube.com/watch?v=MeL-DXslNKs.

John Downer Productions. "Robotic Spy Puppy Meets Wild Dogs." YouTube. May 22, 2017. https://www.youtube.com/watch?v=bfJUciAxxyo.

Kayser, Mark. "Outlawing Trail Cameras Coming to Your State Soon." *Outdoor Life* (blog). April 14, 2010. https://www.outdoorlife.com/blogs/big-buck-zone/2010/04/outlawing-trail-cameras-coming-your-state-soon/.

Kelly, Tara. "Austrian Politician Photographed Having Sex in Forest by Hidden Wildlife Camera." *Huffington Post* (website). June 16, 2012. https://www.huffingtonpost.com/2012/06/16/austrian-politician-photographed-sex-camera_n_1602503.html.

King, Amber, processer. *Gregory Brothers Papers.* Chicago Academy of Sciences/Peggy Notebaert Nature Museum, Finding Aid. March 2017. https://naturemuseum.org/wp-content/uploads/2020/05/FindingAid_Gregory-Brothers.pdf.

Kingtrout435 [pseud.]. "Trail Camera-Ants." Bowsite (forum). June 4, 2008. https://forums.bowsite.com/TF/bgforums/thread.cfm?threadid=351339&forum=2.

Klowsowski, Thorin, and Rebecca Fishbein. "Know Your Rights: Photography in Public." Lifehacker (website). September 25, 2019. https://lifehacker.com/5912250/know-your-rights-photography-in-public.

Lehmberg, Judy. "A Fox and a Badger Fight in Yellowstone—The Whole Story." YouTube. February 24, 2015. https://www.youtube.com/watch?v=Wc8tidOjhUY&t=11s.

Lezappen [pseud.]. "Sped Up Timelapse Footage of Giant Hornet Queen Building Nest." YouTube. February 26, 2017. https://www.youtube.com/watch?v=V3sj6edu3Go&t=13s.

Lowrey, Sassafras. "Conservation Dogs Sniff Out Wildlife." American Kennel Club (website). March 3, 2020. https://www.akc.org/expert-advice/news/conservation-dogs-sniff-out-wildlife-diseases-more/.

Mikkelson, David. "Does Vanilla Flavoring Come from Beaver Anal Secretions?" Snopes (website). Accessed December 7, 2022. https://www.snopes.com/food/ingredient/castoreum.asp.

Mills, Zach. "Ethics of Wildlife Photography." The Wildlife Collective (website). February 2, 2022. https://thewildlifecollective.com/ethics-of-wildlife-photography/.

Milojevic, M. "*Castoreum*, Castor fiber (Castoridae)." Fragrantica (website). Accessed December 7, 2022. https://www.fragrantica.com/notes/Castoreum-102.html.

Naish, Darren. "The Mystery Mammal of Kayan Mentarang." *Scientific American* (blog). February 25, 2013. https://blogs.scientificamerican.com/tetrapod-zoology/mystery-mammal-of-kayan-mentarang/.

NANPA Ethics Committee. "Ethics." North American Nature Photography Association (website). Accessed December 2, 2022. https://nanpa.org/why-nanpa/ethics/.

Nat Geo Live. "Steve Winter: On the Trail of the Tiger." YouTube. March 14, 2012. https://www.youtube.com/watch?v=hVOHKWnflTY.

National Park Service. "Citizen Science." NPS (website). Accessed December 6, 2022. https://www.nps.gov/subjects/citizenscience/index.htm.

National Park Service. "Filming and Still Photography Permits." NPS (website). Accessed December 6, 2022. https://www.nps.gov/aboutus/news/commercial-film-and-photo-permits.htm.

National Park Service. "Scientific Research and Collections Permits." NPS (website). Accessed December 6, 2022. https://www.nps.gov/seki/planyourvisit/scientific-research-and-collection-permits.htm.

Nelson, Stephen. "Flooding Hazards, Prediction and Human Intervention." *Natural Disasters EENS 3050*. Tulane University Department of Earth and Environmental Sciences (website). Last updated October 19, 2015. https://www2.tulane.edu/~sanelson/Natural_Disasters/floodhaz.htm.

News Corp Australia Network. "Beaver Anal Secretions Used as Food Flavouring, Swedish Food Agency Confirms." *News Corp Australia Network* (website). September 16, 2013. https://www.news.com.au/lifestyle/food/beaver-anal-secretions-used-as-food-flavouring-swedish-food-agency-confirms/news-story/86b325da2013bf9a948682c8de2b1f8f.

Nichols, Michael. "Full Disclosure." Photocrati (website). January 30, 2010. https://www.photocrati.com/full-disclosure/.

North Carolina Museum of Natural Sciences. "Coyote Finds Den of Red Fox Pups." YouTube. July 1, 2015. https://www.youtube.com/watch?v=NqBrWMGtFLI.

Opstad, Nessa. "Can Flash Photography Damage Eyes? [Explained It]." NESOP (website). Accessed December 7, 2022. https://nesop.com/can-flash-photography-damage-eyes-explained-it/.

Parliament of Owls. "Daytime Cats! The Lion King Scraping in Broad Daylight." YouTube. April 29, 2014. https://www.youtube.com/watch?v=j4tO96cMCrc.

Pearson, Gwen. "CDC Warning on Misuse of Pest Strips." Wired.com. January 29, 2014. https://www.wired.com/2014/01/cdc-warning-misuse-pest-strips/.

Peetre [pseud.]. "Paper Wasps Making a Nest." YouTube. July 10, 2012. https://www.youtube.com/watch?v=sxVHSSbGW4Q.

Pjdoc1 [pseud.]. "Ants vs. Trail Cams." Hunting Indiana (forum). June 19, 2013. https://hunt-indiana.com/thread/47172/ants-trail-cams.

Prior, John. "Best Moose Documentary Ever." YouTube. March 6, 2017. https://www.youtube.com/watch?v=2UZ-pOvnAL_M&t=2766s.

Project Wet Foundation. "Discover Yellowstone." USGS.gov. Accessed February 12, 2023. https://www.projectwet.org/sites/default/files/content/documents/discover-yellowstone-indicator-game.pdf.

Quinn, Gene. "Sample DMCA Take Down Letter." *IPWatchdog* (blog). July 6, 2009. https://ipwatchdog.com/2009/07/06/sample-dmca-take-down-letter/id=4501/.

Rattlerjen [pseud.]. "How Scent and Airflow Works." *Hound and the Found* (blog). February 22, 2012. https://houndandthefound.wordpress.com/2012/02/22/how-scent-and-airflow-works/.

Ritchie, Bryan. "How to Upload Way Points and Tracks to Google Earth." YouTube. September 7, 2016. https://youtu.be/O86jBT6h25A.

Roberts, Claude. "Sniffer Dogs: The World of K9 Scent Detection." *Dog Learner* (blog). July 24, 2020. https://dog-learner.webflow.io/articles/sniffer-dogs-the-world-of-k9-scent-detection.

ROD917 [pseud.]. "Electronic Predator Call." AutoDesk Instructables (website). Accessed December 2, 2022. https://www.instructables.com/Electronic-Predator-Call/.

Rohde, Ashley. "*Brachylagus idahoensis* pygmy rabbit." Animal Diversity Web (website). Accessed November 30, 2022. https://animaldiversity.org/accounts/Brachylagus_idahoensis/.

Rondeau, Emmanuel. "Professional Camera Trapping." Emmanuel Rondeau.com. Accessed December 6, 2022. https://emmanuelrondeau.com/publications/pro-camera-traps/.

Sample, Ian. "Wildlife Documentaries Infringe Animals' Privacy, Says Report." *The Guardian* (website). April 29, 2010. https://www.theguardian.com/environment/2010/apr/29/wildlife-films-infringe-privacy.

Savage, Emily. "The Science of Scent." *Trailie Paws For Thought* (blog). September 19, 2022. https://www.trailiepawsforthought.com/post/the-science-of-scent-part-one.

Schneier, Bruce. "Human/Bear Security Trade-Off." *Schneier on Security* (blog). August 18, 2006. https://www.schneier.com/blog/archives/2006/08/security_is_a_t.html.

Schroyen, Gary. "Sony 3N Video Build with LED Lighting." *Camtrapper* (blog). September 8, 2016. https://camtrapper.com/viewtopic.php?f=5&t=10811.

Skelton, Jeremi. "Not a Good Feeling." *Camtrapper* (blog). December 13, 2014. https://camtrapper.com/viewtopic.php?t=9185#p73096.

Smartt, Casey. "Bunker Hill Wireless Build." *Camtrapper* (blog). March 16, 2016. https://camtrapper.com/viewtopic.php?f=113&t=1040.

Smartt, Casey. "Down-n-Dirty Camera Skid." *Camtrapper* (blog). September 7, 2014. https://camtrapper.com/viewtopic.php?f=113&t=8880&p=69851&hilit=quick+and+dirty#p69851.

Smartt, Casey. "New AIR Build." *Camtrapper* (blog). February 11, 2018. https://camtrapper.com/viewtopic.php?f=113&t=11702#p96042.

Smartt, Casey. "Wireless Infrared Beam Break Units." *Camtrapper* (blog). January 20, 2018. https://camtrapper.com/viewtopic.php?f=113&t=11667#p95703.

Smithsonian Insider. "New Species of Giant Rat Discovered in Crater of Volcano in Papua New Guinea." *Smithsonian Insider* (website). September 9, 2009. https://insider.si.edu/2009/09/new-species-of-giant-rat-discovered-in-crater-of-volcano-in-papua-new-guinea/.

Smoller, Louis. "Using Orphan Work." *Artrepreneur* (blog). March 11, 2021. https://artrepreneur.com/journal/using-orphan-work/.

Snow Leopard Conservancy. "Foxlights." Snow Leopard Conservancy (website). Accessed December 1, 2022. https://snowleopardconservancy.org/foxlights-video/.

Staff Writer. "Stolen Trail Camera Sent Back Photos of Suspects." *Lincoln Times-News* (website). January 9, 2015. https://www.lincolntimesnews.com/local-news/top/stolen-trail-camera-sent-back-photos-of-suspects/article_3fcd2f81-a737-5650-a9ff-1bf6cc1e2bbb.html.

State of California. "Rodenticides: Rodenticides Can Harm Wildlife." California Department of Fish and Wildlife (website). Accessed December 1, 2022. https://www.wildlife.ca.gov/Living-with-Wildlife/Rodenticides.

Stolen Camera Finder. "How It Works." StolenCameraFinder.com (website). Accessed December 5, 2022. https://www.stolencamerafinder.com/howitworks.jsp.

Stone Cabin Photos. "Bears and Gray Fox Dig into Wasp Nest." YouTube. August 28, 2015. https://www.youtube.com/watch?v=tAFnqcjdzTw.

Story Trender. "Animals in Mirrors Hilarious Reactions." YouTube. June 1, 2015. https://www.youtube.com/watch?v=GaMylwohL14.

Swedish Environmental Protection Agency. "Game Camera for Private Use." Natur Vards Verket [Swedish Environmental Protection Agency] (website). Accessed April 6, 2023. https://www.naturvardsverket.se/vagledning-och-stod/jakt-och-vilt/viltkamera/.

Taursir [pseud.]. "Squirrel Daylight Robbery I." YouTube. March 15, 2007. https://www.youtube.com/watch?v=n-mQXu4cQRao.

TCILPart3. "The Burrowers: Animals Underground—Episode 1." YouTube. August 16, 2013. https://www.youtube.com/watch?v=W76EDYEF3b8.

TCILPart3. "The Burrowers: Animals Underground—Episode 2." YouTube. August 23, 2013. https://www.youtube.com/watch?v=9qL7hnDcyWA.

TCILPart3. "The Burrowers: Animals Underground—Episode 3." YouTube. August 30, 2013. https://www.youtube.com/watch?v=Z_g61gulKi4.

The Local. "Beaver Butt Secretion Good for Baking: Agency." *The Local* (website). September 15, 2013. https://www.thelocal.se/20130915/50250.

US Fish and Wildlife Service. "Laws, Agreements, and Treaties." FWS (website). Accessed December 6,2022. https://www.fws.gov/library/categories/laws.

US Fish and Wildlife Service. "Species." FWS (website). Accessed December 6, 2022. https://fws.gov/species.

US Fish and Wildlife. "Volunteer Opportunities." FWS (website). Accessed December 6, 2022. https://www.fws.gov/volunteer-opportunity.

Vedantu. "Jacobson's Organ." Vedantu.com (website). Accessed December 2, 2022. https://www.vedantu.com/biology/jacobsons-organ.

Viral Hogs. "Red Foxes enjoy Elk Carcass." YouTube. March 3, 2017. https://www.youtube.com/watch?v=ezt1bqLy-qpo.

Wangchuk, Rinchen Norbu. "How a Forest Officer Clicked India's First Photo of a Tiger in the Wild," *The Better India* (website). June 16, 2022. https://www.thebetterindia.com/288507/how-forest-officer-frederick-walter-took-indias-first-tiger-pic/.

Water Science School. "Snowmelt Runoff and the Water Cycle." USGS (website). June 12, 2019. https://water.usgs.gov/edu/watercyclesnowmelt.html.

Wemmer, Chris. "30-Year-Old Possum Picture." *Camera Trap Codger* (blog). February 12, 2009. https://cameratrapcodger.blogspot.com/2009/02/30-year-old-possum-picture.html.

Wemmer, Chris. "Aerial Dogfights of Spring." *Camera Trap Codger* (blog). March 24, 2007. https://cameratrapcodger.blogspot.com/2007/03/aerial-dogfights-of-spring.html.

Wemmer, Chris. "Bat Strafes Owl." YouTube. July 23, 2003. https://www.youtube.com/watch?v=YYm9Il75dmU.

Wemmer, Chris. "Bears on the Puma Kill." *Camera Trap Codger* (blog). December 13, 2017. https://cameratrapcodger.blogspot.com/2017/12/bears-on-puma-kill.html.

Wemmer, Chris. "Bears vs. the Squeaker." *Camera Trap Codger* (blog). July 20, 2013. https://cameratrapcodger.blogspot.com/2013/07/bears-vs-squeaker.html.

Wemmer, Chris. "Camera Trap Pioneers: Nicholas Smythe." *Camera Trap Codger* (blog). January 31, 2010. http://cameratrapcodger.blogspot.com/2010/01/camera-trap-pioneers-nicholas-smythe.html.

Wemmer, Chris. "Creature Comfort." YouTube. August 8, 2013. https://www.youtube.com/watch?v=vcmjqsUM_x8.

Wemmer, Chris. "Denning Kit Foxes." *Camera Trap Codger* (blog). August 18, 2013. https://cameratrapcodger.blogspot.com/2013/08/denning-kit-foxes.html.

Wemmer, Chris. "Hard-to-Get Tree Vole." *Camera Trap Codger* (blog). January 15, 2012. https://cameratrapcodger.blogspot.com/2012/01/hard-to-get-tree-vole.html.

Wemmer, Chris. "On Camouflaged Trail Cams." *Camera Trap Codger* (blog). January 4, 2011. https://cameratrapcodger.blogspot.com/2011/01/on-camouflaged-trail-cams.html.

Wemmer, Chris. "Spooky Cameras and Squirrelly Neophobia." *Camera Trap Codger* (blog). January 8, 2012. https://cameratrapcodger.blogspot.com/2012/01/spooky-cameras-and-squirrelly-neophobia.html.

Wemmer, Chris. "The Not-So-Giant Civet of Sulawesi—Part 1." *Camera Trap Codger* (blog). February 19, 2009. https://cameratrapcodger.blogspot.com/2009/02/not-so-giant-civet-of-sulawesi-part-1_19.html.

Wemmer, Chris. "The Restless Haystack." *Camera Trap Codger* (blog). October 23, 2011. https://cameratrapcodger.blogspot.com/2011/10/restless-haystack.html.

Wemmer, Chris. "The Submariner's Solution." *Camera Trap Codger* (blog). May 12, 2007. https://cameratrapcodger.blogspot.com/2007/05/submariners-solution.html.

Wemmer, Chris. "Video Vision in the Tunnel—Part 2." *Camera Trap Codger* (blog). November 19, 2012. https://cameratrapcodger.blogspot.com/2012/11/video-vision-in-tunnel-part-2.html.

Wemmer, Chris. "Wet and Thirsty Raptors." YouTube. September 25, 2013. https://www.youtube.com/watch?v=GK-JkAAgtWTE.

Wikipedia. "Carl Georg Schillings." Wikimedia Foundation. Last updated June 29, 2022. https://de.wikipedia.org/wiki/Carl_Georg_Schillings.

Wikipedia. "Eye-Fi." Wikimedia Foundation. Last updated January 21, 2023. https://en.wikipedia.org/wiki/Eye-Fi.

Wildfelids. "Mountain Lion Making a Territorial Scrape." YouTube. July 20, 2016. https://www.youtube.com/watch?v=Q7huAH_XzaQ.

Wildlife Conservation Society. "In Iran, Camera Traps Reveal Rare Asiatic Cheetahs." *ScienceDaily* (website). Accessed November 30, 2022. https://www.sciencedaily.com/releases/2005/09/050912075519.htm.

Yang, Brian. "What Causes the Red Eye Effect?" *Yale Scientific* (website). May 12, 2011. https://www.yalescientific.org/2011/05/what-causes-the-red-eye-effect/.

ZovTV. "Bears vs. Bees." YouTube. April 29, 2009. https://www.youtube.com/watch?v=D-A44El_DMU.

Recommended Reading

There are a multitude of resources for those interested in camera trapping, wildlife photography, and their practitioners. Some of my favorites are listed here.

Bevis, John. *The Keartons: Inventing Nature Photography*. Devon, UK: Uniform Books, 2016.

Bright, Michael. *100 Years of Wildlife*. London: BBC Books, 2007.

Champion, Frederick Walter. *With a Camera in Tiger-Land*. Garden City, NY: Doubleday, Doran & Company, 1928.

Champion, Frederick Walter. *The Jungle in Sunlight and Shadow*. New York: Charles Scribner's Sons, 1934.

Champion, Frederick Walter, and James Champion. *Tripwire for a Tiger: Selected Works of F. W. Champion*. Chennai: Rainfed Books, 2012.

Childs, Jack L., and Anna Mary Childs. *Ambushed on the Jaguar Trail: Hidden Cameras on the Mexican Border*. Tucson, AZ: Rio Nuevo Publishers, 2008.

Clothier, Zack. *The Ultimate Guide to DSLR Camera Trapping*. Self-published, 2022. PDF. https://www.zackclothier-photography.com/dslr-camera-trapping-ebook.

Elbroch, Mark. *Mammal Tracks and Sign: A Guide to North American Species*. Mechanicsburg, PA: Stackpole Books, 2003.

Gackstetter, Chuck. *Trail Camera Trails of Lincoln County, Minnesota*. N.p: Chuck Gackstetter, 2012.

Gregory, Tappan. *Deer at Night in the North Woods*. Springfield, IL: Charles C. Thomas, 1930.

Gregory, Tappan. *Eyes in the Night*. New York: Thomas Y Crowell Co., 1939.

Gregory, Tappan. *Nature Photography at Night: Museum Pictorial No. 14*. Denver, CO: Denver Museum of Natural History, 1957.

Harris, Melissa, and Michael Nichols. *A Wild Life: A Visual Biography of Photographer Michael Nichols*. New York: Aperture, Aperture Foundation, 2017.

Joubert, Dereck, and Beverly. *The Africa Diaries: An Illustrated Memoir of Life in the Bush*. Washington, DC: Adventure Press, National Geographic, 2001.

Kavanaugh, James. *Animal Tracks: A Folding Pocket Guide to the Tracks and Signs of Familiar North American Species (A Pocket Naturalist Guide)*. Dunedin, FL: Waterford Press, 2000.

Kays, Roland. *Candid Creatures: How Camera Traps Reveal the Mysteries of Nature*. Baltimore: Johns Hopkins University Press, 2016.

Kearton, Richard. *A Naturalist's Pilgrimage*. New York: Houghton Mifflin, 1926.

Kekule, L. Bruce. *Thailand's Natural Heritage: A Look at Some of the Rarest Animals in the Kingdom*. Bangkok: WKT Publishing Company, 2006.

Meek, Paul, and Peter Fleming, eds. *Camera Trapping: Wildlife, Management and Research*. Clayton South, Victoria, Australia: CSIRO Publishing, 2014.

Morrissey, Rory. *Oliver G. Pike 1877–1963: Leighton Buzzard Naturalist, Author, Photographer*. Bedfordshire: Leighton Buzzard and District Archaeological and Historical Society, 2013.

Murie, Olaus Johan, and Mark Elbroch. *Peterson Field Guide to Animal Tracks*. Boston: Houghton Mifflin Company, 2005.

Nesbit, William. *How to Hunt with the Camera: A Complete Guide to All Forms of Outdoor Photography*. New York: Dutton & Company, 1926.

O'Connell, Alan F., James D. Nichols, and K. Ullas Karanth, eds. *Camera Traps in Animal Ecology: Methods and Analyses*. New York: Springer Publishing, 2011.

Palmer, Chris. *Shooting in the Wild: An Insider's Account of Making Movies in the Animal Kingdom*. San Francisco: Sierra Club Books, 2010.

Palmer, Chris. *Confessions of a Wildlife Filmmaker: The Challenges of Staying Honest in an Industry Where Ratings Are King*. Philadelphia: Bluefield Publishing, 2015.

Pesaturo, Janet. *Camera Trapping Guide: Tracks, Sign, and Behavior of Eastern Wildlife*. Harrisburg, PA: Stackpole Books, 2018.

Rhoades, Kevin G. *Wildlife Stalker: Days in the Life of Filmmaker Bob Landis*. Missoula, MT: Five Valleys Press, 2011.

Root, Alan. *Ivory, Apes and Peacocks: Animals, Adventure and Discovery in the Wild Places of Africa*. London: Vintage, Random House, 2013.

Rovero, Francesco, and Fridolin Zimmermann, eds. *Camera Trapping for Wildlife Research*. Exeter, UK: Pelagic Publishing, 2016.

Seal, Mark. *Wildflower: An Extraordinary Life and Mysterious Death in Africa*. New York: Random House, 2010.

Schillings, Carl Georg. *With Flashlight and Rifle: Photographing with Flashlight at Night the Wild Animal World of Equatorial Africa*. New York: Harper & Brothers, 1905.

Shiras, George. *Hunting Wildlife with Camera and Flashlight*. Vols. 1 and 2. 3rd ed. Washington, DC: National Geographic Society, 1936.

Tanner, Dawn. *Taking Action Opportunities: Connecting Kids to Wildlife with Trail Cameras*. Minnesota: State of Minnesota, Department of Natural Resources, 2009.

Young, Susan. *CCTV for Wildlife Monitoring: An Introduction*. Exeter, UK: Pelagic Publishing, 2016.

About Chris Wemmer

Photo by Bea Heiloway

CHRIS WEMMER grew up in San Francisco and the Santa Cruz Mountains of California. After earning his bachelor's and master's degrees in biology from San Francisco State University and his PhD in zoology with a major in ethology and minors in ecology and physiology from the University of Maryland, Wemmer worked at the Chicago Zoological Society as curator of small mammals and large carnivores for two years. In 1974 he was hired by the Smithsonian Institution's National Zoological Park, where he was tasked with managing the newly acquired Conservation and Research Center, now known as the Smithsonian Conservation Biology Institute, located in Front Royal, Virginia. Following a year and a half assignment in Washington, DC, as the zoo's acting director, Wemmer continued the long journey of developing an internationally recognized program in conservation science. Over the span of thirty years, the center's programs in captive breeding and husbandry of endangered species grew to include robust initiatives in research, environmental education, and international training in wildlife conservation. Despite his administrative duties, Wemmer has always remained a boy naturalist at heart.

About BRG Scientific

BRG SCIENTIFIC was founded to advance scholarly works of the highest intellectual quality. We place priority on books showcasing the topics of ecology, natural history, evolution, animal behavior, and comparative biology written by vetted colleagues. Our team is driven by excellence of scholarship, not by volume of sales. We strive to reach technical audiences as well as curious, non-technical readers wishing to expand the horizons of their knowledge. We recognize that rigorous, well-researched science can also be pleasing to the eye and soul and that beautiful and informative graphic illustrations complement and enhance the written narrative.

—Current BRG Scientific Releases—

CARVING NATURE AT ITS JOINTS

Theodore I. Grand & Carol E. Underwood

—New Releases by BRG Scientific Coming Soon—

NAPATREE POINT: A DECADE OF STEWARDSHIP

Peter August, Janice Sassi & Daniel T. Cole

SUBSTANCE OVER STYLE: A FIELD GUIDE TO LEADERSHIP IN HIGHER EDUCATION

Donald DeHayes

MAJESTY, POWER & RESILIENCE: A PHOTOGRAPHIC ESSAY OF THE AMERICAN WEST

Vince Warren

SPIRIT OF NATURE IN NORTHERN NEW MEXICO

Brian Miller

BRG SCIENTIFIC

Index

The letter "f" following a page number denotes an illustration.

C

D

E

F

G

H

I

J

K

L

M

S

T

U

V

W

Y

www.ingramcontent.com/pod-product-compliance
Lightning Source LLC
LaVergne TN
LVRC101545220926
848486LV00041B/380

* 9 7 9 8 9 8 6 2 8 2 1 2 1 *